Lecture Notes in Electrical Engineering

Volume 420

About this Series

"Lecture Notes in Electrical Engineering (LNEE)" is a book series which reports the latest research and developments in Electrical Engineering, namely:

- Communication, Networks, and Information Theory
- Computer Engineering
- Signal, Image, Speech and Information Processing
- Circuits and Systems
- Bioengineering

LNEE publishes authored monographs and contributed volumes which present cutting edge research information as well as new perspectives on classical fields, while maintaining Springer's high standards of academic excellence. Also considered for publication are lecture materials, proceedings, and other related materials of exceptionally high quality and interest. The subject matter should be original and timely, reporting the latest research and developments in all areas of electrical engineering.

The audience for the books in LNEE consists of advanced level students, researchers, and industry professionals working at the forefront of their fields. Much like Springer's other Lecture Notes series, LNEE will be distributed through Springer's print and electronic publishing channels.

More information about this series at http://www.springer.com/series/7818

Electronic Navigation Research Institute
Editor

Air Traffic Management and Systems II

Selected Papers of the 4th ENRI International Workshop, 2015

Springer

Editor
Electronic Navigation Research Institute
(ENRI)
Tokyo, Japan

ISSN 1876-1100 ISSN 1876-1119 (electronic)
Lecture Notes in Electrical Engineering
ISBN 978-4-431-56421-8 ISBN 978-4-431-56423-2 (eBook)
DOI 10.1007/978-4-431-56423-2

Library of Congress Control Number: 2017930154

Printed on acid-free paper

This Springer imprint is published by Springer Nature
The registered company is Springer Japan KK
The registered company address is: Chiyoda First Bldg. East, 3-8-1 Nishi-Kanda, Chiyoda-ku, Tokyo 101-0065, Japan

Preface

This is the second book edited by the Electronic Navigation Research Institute (ENRI) after its workshops. ENRI is a national laboratory in Japan specialized in air traffic management (ATM) and communication, navigation, and surveillance (CNS) for aviation. ENRI organizes workshops titled "ENRI International Workshop on ATM/CNS (EIWAC)" to contribute to the development of civil aviation by exchanging and sharing updated information about ATM/CNS in the world. The fourth workshop, EIWAC2015, was held in November 2015 in Tokyo.

It will be a great honor to ENRI if EIWAC meetings contribute opportunities for discussions on future skies. Many keynote speakers in the meetings indicated the importance of conferences like EIWAC to share views on future air transport operations and to carry out harmonized, coordinated actions.

The topics in this book include those of presentations in EIWAC2015 from the keynote speeches, reports for technical interchanges, and academic discussions. The technical papers reported great progress after EIWAC2013 in many areas such as mathematical models of aircraft trajectory and optimization of air traffic.

The contents of this book are as follows. An introduction to EIWAC2015 is provided in Part I. The chapters on trajectory planning and optimization are in Part II, and the chapters on optimization of air traffic are in Part III. The chapters on enablers are in Part IV. More than 20 papers in the area of CNS were presented at EIWAC2015 while only two papers are found in Part IV. Most contents of the papers not in this book will appear in other publications such as related project reports or as academic papers.

The chapters in the book were selected and passed through a peer-review process as papers in academic journals do. Technical topics were selected from the proceedings presented in the technical sessions for EIWAC2015. The presentations were appreciated by onsite auditors in the ATM and CNS areas for their usefulness. In addition, they were recognized by reviewers for their significant contribution to new proposals, methods, experience, or surveys as they summarized essential papers in related areas. Indirect interactions between reviewers and authors made the papers clearer and more useful. After these selection processes, the articles were compiled to constitute the chapters of this book.

Editors for the book are grateful to the technical reviewers for their valuable contributions. It can be difficult to find technical reviewers without the help of a large academic society. Even so, the editors, with their determination to contribute to harmonizing ATM in the future, were fortunate to find many capable technical reviewers for this work.

I hope the book will become accessible for the readers around the world who could not participate in EIWAC2015. I also hope the book will be well received by students who are deciding on their major area of endeavor for the future.

I would like to express my deep appreciation to EIWAC2015 Technical Program Committee members and to the associate editor, who volunteered their contributions to revise the editorial policy, and for their special support in the reviewing process. They are listed in this book with special thanks.

Tokyo, Japan Shigeru Ozeki

Contents

Committee Members' List

EIWAC2015 Technical Program Committee

Editor in Chief

Shigeru Ozeki [Electronic Navigation Research Institute (ENRI), Japan]

Editors

Vu Nguyen Duong [Viet Nam National University, Ho Chi Minh City (VNU-HCM), Viet Nam]

Jean-Marc Loscos [Direction des Services de la Navigation Aérienne (DSNA), France]

William C. Johnson [National Aeronautics and Space Administration (NASA), U.S.A.]

Masatoshi Harigae [Japan Aerospace Exploration Agency (JAXA), Japan]

Takeshi Tsuchiya [The University of Tokyo, Japan]

Keisuke Matsunaga [ENRI, Japan]

Susumu Saito [ENRI, Japan]

Takuya Otsuyama [ENRI, Japan]

Atsushi Senoguchi [ENRI, Japan]

Eri Itoh [ENRI, Japan]

Shun-ichi Futatsumori [ENRI, Japan]

Associate Editor

Jon Holbrook [NASA, U.S.A.]

Secretaries

Akiko Kohmura [ENRI, Japan]

Atsushi Kezuka [ENRI, Japan]

Atsushi Senoguchi [ENRI, Japan]

Part I
Introduction

Introduction to the Fourth ENRI International Workshop on ATM/CNS (EIWAC 2015)

Kazuo Yamamoto

Abstract ENRI organized the fourth ENRI International Workshop on ATM/CNS (EIWAC 2015) to share comprehensive information on the latest ATM/CNS technologies and operations among EIWAC participants. In this chapter, the overview of EIWACs, summaries of each keynote speech and invited speech are presented. The discussion in the EIWAC has demonstrated the problems that present aviation systems are facing and the perspective for future global aviation systems. Harmonization between current and future ATM/CNS systems, R&D and implementation, and different modernization plans are also discussed.

Keywords ENRI • ATM • CNS • EIWAC • Standardization • Global Air Navigation Plan

1 Introduction

Electronic Navigation Research Institute (ENRI) is an affiliate of the Ministry of Land Infrastructure, Transport and Tourism (MLIT) Japan. ENRI has been conducting research, development, and test on electronic navigation systems for almost half a century. ENRI is now an only institute in Japan specializing in air traffic management (ATM), communication, navigation, and surveillance (CNS) for aviation.

Air traffic is increasing steadily in the world. And mitigation of congestion and reduction of environmental impact, while keeping safety and efficiency, have become world common interest. Especially in the Asia Pacific region, demand for increasing air traffic capacity, efficiency, and safety has been serious because recent growth rate of air traffic in the region is the highest in the world. ENRI is then conducting R&D to respond to the demand and to provide results timely for national, regional, and global aviation systems' improvement.

K. Yamamoto (✉)
ENRI, Tokyo, 185-0002, Japan
e-mail: k.yamamoto.2246@gmail.com

Electronic Navigation Research Institute, *Air Traffic Management and Systems II*,
Lecture Notes in Electrical Engineering 420, DOI 10.1007/978-4-431-56423-2_1

In addition to the R&D activities, ENRI is now expected to contribute to harmonization and standardization on the current and emerging ATM/CNS technologies and operations, which will lead to fully harmonized global aviation systems with modern performance-based technologies and procedures. However, it is not necessarily easy for the aviation community in the Asia Pacific region to share comprehensive information on the latest ATM/CNS technologies and operations because of the region's different traffic policies and CNS/ATM capabilities.

Thus, ENRI decided to organize an international workshop discussing ATM/CNS technologies, operations, and Asian sky improvement. This workshop was named "ENRI International Workshop on ATM and CNS (EIWAC)." The first workshop was held in 2009, and the newest one (EIWAC 2015) was in November 2015 in Tokyo.

This book is published for the purpose of providing distinguished topics presented and discussed in the EIWAC 2015. In this chapter, the overview of EIWACs is described first. Then the speaker, title, and summary of each keynote addressed in the plenary session are presented. And the invited talks on R&D for global harmonization are summarized. Papers selected and reviewed from the contributions will be provided in the following chapters.

2 Overview of EIWACs

In March 2009, ENRI organized the ENRI International Workshop on ATM and CNS [1] in Tokyo under the theme "Towards Future ATM/CNS." It was the first international workshop on ATM/CNS in Japan. Twenty-two presentations including four keynote speeches were provided in the EIWAC. The second EIWAC [2] was held at Tokyo Akihabara in 2010 with the theme "Safety, efficiency and environment." Forty-five presentations including seven keynote speeches in 19 sessions were performed. A panel session was added to discuss future automated ATM. The third EIWAC [3] was held at Tokyo Odaiba in 2013 with the theme "Drafting future sky." Forty-six presentations including nine keynote speeches in 17 sessions were performed. A panel session was opened under the title "Future ATM: Centralized, decentralized or best mixed?" The fourth EIWAC [4] (EIWAC 2015) was held at Tokyo Ryogoku on November 17–19, 2015 under the theme "Global Harmonization for Future Sky." This theme came from the recognition that harmonization is indispensable for successful upgrade of global ATM/CNS systems between current and emerging technologies, R&D and implementation, and among different ATM/CNS capabilities.

Table 1 is a brief summary comparing the third and fourth EIWACs. The summaries about the first to third EIWACs were presented in the last publication [5]. The table shows that in EIWAC 2015, the number of keynote and invited speeches increased and the more speakers with different backgrounds expanded the diversity of the topics, which will be very important to discuss global ATM

Table 1 Summary of the EIWAC2013 and 2015

Serial number	3	4
Name	EIWAC 2013	EIWAC 2015
Date, year	February 19–21, 2013	November 17–19, 2015
Venue	Odaiba Miraikan hall, Tokyo	Ryogoku KFC Hall & Rooms, Tokyo
Theme	Drafting future sky	Global Harmonization for Future Sky
Keynote and invited speakers	9	13
Panel/special sessions	Panel Session: "Future ATM: Centralized, decentralized, or best mixed" by 4 panelists	Asia Session: "Asian R&D Interchange" by 5 Asian specialists
Total sessions	17	18
Total presentations	46 including 33 foreign speakers	70 including 40 foreign speakers
Participants	540 including 80 foreign guests	744 including 174 foreign guests
The States of participants	13	17
Proceedings	The third ENRI International Workshop on ATM/CNS, Feb. 2013	The fourth ENRI International Workshop on ATM/CNS, Nov. 2015
The number of organizing/programming committee members and their affiliations	20, from ENRI, University of Tokyo (UoT), Japan Aerospace Exploration Agency (JAXA), Japan Civil Aviation Bureau (JCAB), Direction des Services de la Navigation Aérienne (DSNA) and Korea Aerospace Research Institute (KARI)	23, from ENRI, UoT, JAXA, DSNA, Vietnam National University (VNU) and National Aeronautics and Space Administration (NASA)
Supporters	13	16

harmonization. The nationality of the participants also increased in number. The quality of presentations has been improved from one event to next by enhancing the reviewing process to submitted contributions with each event. In the EIWAC 2015, three international specialists from France, the USA, and Vietnam joined the reviewing. This careful and precise reviewing process enabled us the publication of this "Air Traffic Management and Systems" series.

Additional features of the EIWAC 2015 were:

- EIWAC 2015 welcomed the largest number of keynote and technical speakers in the EIWAC series.
- The EIWAC became the first Asian event for the EUROCAE representative to make a speech.

- The largest number of students and young researchers joined the EIWAC 2015 and activated the discussion thanks to the encouragement of professors with whom ENRI is now collaborating.
- Remarkable increase in the number of participants from Asian countries especially from China and Singapore.

Table 1 shows that more and more people in the world have been interested in the EIWAC with each event. The topics in the EIWAC became more and more updated, refined, and diversified. Therefore, we believe that EIWAC has become one of the most active and attractive symposiums about ATM/CNS and related topics in the world.

3 Keynote Speeches

Keynote speakers and presentations at the plenary sessions in EIWAC2015 are summarized in this section. The presentation slides are available at ENRI website [4].

3.1 Hitoshi Ishizaki, "Toward the Realization of Seamless SKY in the Asia Pacific Region"

Hitoshi Ishizaki is the director general of Japan Air Navigation Service (JANS), Civil Aviation Bureau (JCAB), Ministry of Land Infrastructure, Transport and Tourism (MLIT). He presented following three major topics:

- Necessity of realizing seamless sky
- Regional actions to respond to the necessity
- Practical actions in Japan

First, he spoke about the "seams" in the sky, which are created by different operational procedures and CNS capabilities by each country or region. Such "seams" cause different flight separation and procedures in the flight information region (FIR) borders and can bring about heavy congestion, unsafe, and inefficient operation. He then emphasized the importance of removing such "seams" through collective cooperation among aviation communities concerned.

Second, he presented the global or regional cooperation that JCAB is now participating, as ICAO's global activities under the Global Air Navigational Plan (GANP) and as the regional activities based on Regional Air Navigation Plan like Asia/Pacific Seamless ATM Plan. In the Asia/Pacific Plan, priority is placed on PBN (performance-based navigation), network operation, air traffic flow management

(ATFM), aeronautical information management, etc. Three level approaches are being taken as:

1. Regional cooperation such as APANPIRG (Asia Pacific Air Navigation Planning and Implementation Regional Group)
2. Subregional cooperation as Northeast Asia Regional Harmonization Group
3. Bilateral cooperation such as Japan and Korea meeting

He then presented the achievements corresponding to above approaches as:

1. Contribution to ATFM steering group in APANPIRG
2. Establishment of ATM coordination with the Philippines, Hong Kong, Taiwan, and ROK
3. Evaluation and implementation of UPR (User Preferred Route) and DARP (Dynamic Air Route Planning) under JCAB and FAA cooperation

JCAB established a National Air Navigation Plan "CARATS" (Collaborative Actions for Renovation of Air Traffic Systems) to improve the national air traffic systems. Figure 1 shows an example of the CARATS roadmap describing the specific subject, timeline from R&D, preparation to implementation (launching new operation), and decision making. Collaborative actions will be done in all the phases including the decision by all stakeholders. There are 67 roadmaps now to cover all necessary measures for national ATM modernization. The roadmaps are often reviewed and revised.

Third, some examples of practical actions presently taken in Japan were presented:

- R&D and evaluation of trajectory-based operation (TBO) for totally optimized air traffic
- Investigation of SWIM (System Wide Information Management) for smooth and efficient aviation information sharing
- Participation to Mini Global Demonstration (MGD) to exhibit the validity of global flight information sharing

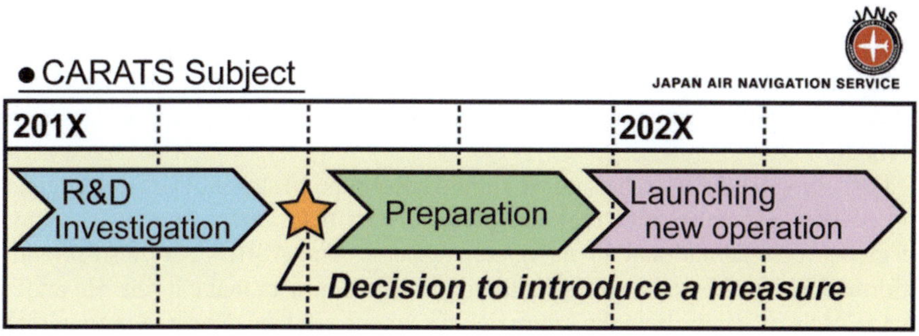

Fig. 1 Example of the CARATS roadmap

Finally, he said that each country should take actions in accordance with their situation and globally/regionally harmonized manner to accommodate future air traffic growth.

3.2 Richard Macfarlane, "ICAO's Global Air Navigation Plan and the Importance of Being Earnest about the ASBUs"

Richard Macfarlane, deputy director of Air Navigation Capacity and Efficiency of ICAO, addressed the concept of Global Air Navigation Plan (GANP), impact of aviation as an economic driver, Aviation System Block Upgrade (ASBU), and selling the ASBUs.

First, ICAO's policy was announced that aims to unite world aviation systems without any countries left behind. He then said that the roadmaps in the GANP provide certainty for aviation system modernization in equipage, for industry, for investment, and for indicating research and development directions. Four trunk roadmaps in the GANP are for airport operation, interoperable systems and data, globally collaborative ATM, and efficient flight tracks. The details of the roadmaps are defined in the ASBU [6].

Second, the potential of aviation was described as an economic driver. Development and implementation of modernized aviation systems may contribute to 2.4 trillion dollars and 58.1 million jobs in the world economy. Future air travel activities of Asian people can be greater than those of Europeans and North Americans because many Asian countries are now enjoying higher economic growth rate than any other regions. He then emphasized that harmonization of safety and development must always be taken into account to make aviation be constant economic driver. And appropriate predictions about future aviation globally, regionally, and nationally must be more and more important for operators and stakeholders.

Third, he explained ASBUs as air navigation building blocks. The ASBUs consist of four blocks. Currently we are in the phase of block 0 and 1. Each block is constantly reviewed by stakeholders from safety, effectiveness, and efficiency points of view. Gap analysis and implementation are regarded as best practices for building the blocks. Necessary actions and measures in each block have been investigated and discussed voluntary when business case and influence diagrams must be taken into account.

For selling the ASBUs, he said we should encourage all stakeholders to recognize the ASBUs as global and regional predictors to help invest in modernization of global/regional/national air navigation systems. The ASBUs may also provide information on who pays, who gets the returns, and how to make it fair. He added that ICAO's estimate about the return on investment will be 3–5 years for operators, 5–10 years for state, 10–15 years in region, and 15 years or more in globe. The accuracy or reliability of this estimation will be examined further.

Finally, he informed that ICAO is providing the Implementation Kits (iKIT) to help us understand the GANP and ASBUs. His presentation indicated the importance of harmonization of policies, operations, and technologies for modernizing aviation systems among states, regions, and the world under the policy of "No countries left behind."

3.3 Michael Standar, "SESAR – The European Approach to Support Global Harmonization and Interoperability"

Michael Standar is the Chief Strategy and External Affairs at the SESAR Joint Undertaking. His presentation was about the need for harmonization and interoperability of world aviation systems. He described the European context first, where 42 EU and non-EU states, 65 control centers, and 1940 aircraft operators are involved. He said that current importance is to achieve Single European Sky and that basic concept includes performance base, a holistic approach, dedicated and evolving legal framework, and involving and supporting stakeholders.

Then, he spoke about the SESAR project. The project comprises three inter-related processes as definition, development, and deployment. In the definition process, SESAR performance targets are defined and described in the European ATM Master Plan. Development work is conducted by all the actors from ground/airborne industries, R&D community, regulators, militaries, airports, service providers, etc. Developed results are evaluated in the deployment process. He added that the greatest importance is that all SEAR members have common performance targets.

The European ATM Master Plan is the roadmap identifying "Essential Operational Changes" for development and implementation leading to full deployment of the SESAR solutions by 2035. The Plan contains following four steps:

1. Performance ambitions (the why)
2. Description and prioritization of technical solutions and mapping to the ICAO Global Context (the what)
3. Deployment scenarios (the where and when)
4. Investment needs and performance gains over time (the how much)

About "the why" in the Master Plan, we should deliver the expected values to airlines and passengers by the deployment of the SESAR solutions. Several examples of the values are:

- 30–40 % reduction in ANS costs per flight
- 5–10 % additional flights at congested airports
- 10–30 % reduction in departure delays
- 5–10 % reduction in CO_2 emissions

Then about "the what," we must observe the transition of ATM visions and industry trends. He added his opinion that automation will be a very important key for future evolution of aviation systems. For "where and when," technology should reach the market at any place on time. And about "how much," he commented that precise estimation of the value of investment must be done for each R&D subject.

He presented some major projects aiming to near future deployment such as SWIM, extended arrival management, ground/ground interoperability, remote tower services, i4D (initial four-dimensional) TBO, etc. Exceeding 60 demonstration projects, 30,000+ flight trials are now being conducted at more than 50 locations to demonstrate the solutions. He then explained the next phase SESAR, SESAR2020. In the SESAR2020, exploratory research, applied and industrial research, and large-scale demonstrations are emphasized to facilitate the transition from R&D to implementation.

In conclusion, he announced that SESAR will support global harmonization and interoperability by bringing all actors in development, operational validations, and alignment of the European ATM Master plan with ICAO GANP/ASBUs.

3.4 Philippe Merlo, "CNS in ATM: The Challenges Ahead"

Philippe Merlo, director of Air Traffic Management, EUROCONTROL started his address from the FANS (Future Air Navigation Systems) report issued by ICAO in 1988. Twenty-seven years after this report, SBAS (satellite-based augmentation system), APV (approach procedures with vertical guidance) approaches, ADS-B (automatic dependent surveillance-broadcast), and WAM (wide area multilateration) have been on implementation phases in European sky. Frequency band compression to 8.33 kHz in VHF COM (communication) mitigated VHF congestion in European sky.

He said, however, that there are many challenges left behind. He presented several practical examples of the challenges as:

- VHF band congestion has been serious again in core Europe.
- Many airborne antennas bring about complicated onboard navigation systems.
- Multiple and excessive radar coverage in core Europe is degrading radio wave environment.
- Different kinds of surveillance systems with different performances bring about confusion in establishing appropriate and optimum surveillance technique.
- Introduction of i4D TBO requires new additional capabilities and procedures in the present CNS systems.

He said that CNS improvement is indispensable to deal with above challenges.

He then presented several drivers that can make us change/improve/modernize present aviation systems as:

- Spectrum efficiency: Review and improvement of frequency allocation is necessary to enable efficient frequency use because legacy spectrum allocation does not suit to current and future radio navigation applications,
- ATM newcomers: RPAS (remotely piloted aircraft system), suborbital flights will be in practical operation soon. We must take into account the features and flight performance of such new comers.
- GNSS failure and backup: New technologies are required to cope with GNSS vulnerability because the impact of GNSS failure will be more serious in future RNP operation.

He explained that SESAR2020 will begin a project about A-PNT (alternative positioning, navigation, and timing) based on DME/DME or other CNS technologies to prepare for the GNSS failure.

He added the potential drivers for change based on current technologies as:

- Software-defined radios
- Microelectromechanical systems
- Multi-static primary surveillance radars
- Sat Com
- Multiband airborne antennas

Most of these are regarded as promising, and many R&D organizations in the world including ENRI are involved in the work to put such technologies into practice.

Finally, he proposed pursuing completely new technologies which can be applied to the present ATM/CNS systems. He concluded that such innovative technologies may cope with ever-increasing air traffic.

3.5 Neil Planzer/Chris Metts, "how Do they all Come together ... SESAR, NextGen and ASBU"

Neil Planzer, the vice president of Airspace Solutions and Air Traffic Management Digital Aviation for Boeing Commercial Airplanes, is leading the development and implementation of the company's ATM strategy and business. His lecture time was shared with Chris Metts, Harris Corporation, to supplement his presentation from a view of "soft path to success."

His first topic was about current and future air traffic system. He said that from aircraft manufacture's point of view, current ATM systems are so constrained and limited in capacity growth that robust, flexible, and globally harmonized ATM systems are critical for success and long-term growth in the ATM business. And future airplane sale is deeply dependent on safe and efficient ATM systems.

He then stressed the importance of holistic approach to ATM improvement as:

- Looking at ATM as a completely integrated, aircraft connected, shared command and control system
- Fully integrating the current and future capabilities of the aircraft
- Utilizing the best CNS capabilities
- Considering the full breadth of the ATM system
- Balancing all stakeholder requirements to ensure that each mission and objective can be accomplished while reducing operation cost
- Creating a more robust, flexible, and seamless ATM system that allows dynamic airspace allocation and more effective and efficient airspace use

He presented metrics for success by fulfilling above holistic approach as:

- Improved safety and security
- Achievement of global interoperability
- Meeting future requirements for civil air traffic systems
- Reduced operating cost
- Transition through mixed fleet operations
- Improved and shared situational awareness between stakeholders
- Capacity and efficiency improvements
- Environmental improvements

Finally, he emphasized that more integrated and holistic approach is indispensable to build globally harmonized ATM system. In particular, we should harmonize ground-based ATM with aircraft capabilities.

Chris Metts, Harris Corporation, had been with FAA for more than 30 years and was former director of FAA Asia Pacific International Office. He, with Neil Planzer of Boeing, addressed the "soft" path to ATM modernization by four important keys as:

1. Engagement of all stakeholders
2. Protecting present mission (operation)
3. Improvement and advancement of safety
4. Addressing culture difference

About the key 1, he said that cooperation, coordination, and early and often engagement with all stakeholders as operators, users, regulators, academics, labor organizers, and global partners are indispensable to achieve seamless sky. And we have to make deliberate and continuous actions by taking into account the complexity of such cooperation and engagement.

Second, he mentioned that early and constant consideration is indispensable about the impact of great ideas and innovations upon critical networks and services supporting the mission and operation of present ATM. The way to success will be harmonization between present and expected future missions, he added.

Third, he emphasized safety is the first always from the views of global standards, global expectations, and global impact, and operations cannot be compromised by any operation or engineering changes.

Fourth, he recommended us to notice culture difference. We have to understand that successive communication is of topmost importance among the stakeholders so that we can introduce appropriate training and innovation for ATM improvement into the nations or regions with different culture and environment.

In summary, he pointed out that deliberate leadership with consistent soft skills is needed beyond innovation, because success of NextGen, SESAR, and ASBU is dependent upon communications, awareness, training, and metrics.

3.6 *Blair Cowles, "The Operators' Perspective on ATM Modernization"*

Blair Cowles is IATA's Asia-Pacific Regional director of safety and flight operations. He first introduced IATA briefly. IATA is a global trade association for the world's airlines with 250 passenger and cargo carriers which carry 84 % of global air traffic. Its major role is to meet its member's needs by demonstrating the tremendous value that aviation creates for global economy. It is now acting for the following key " to" issues:

- Continually improve aviation safety
- Increase value through partnership
- Protect the interests of the industry
- Reduce environmental impact

He described the growth of aviation in Asia Pacific (APAC) in terms of the number of middle-class people. ICAO recognizes that the growth of middle-class people is a key to air travel growth. He said that APAC is expected to enjoy the largest growth rate of middle-class people in the world from 2009 to 2030, which will promote regional industry growth as well. When we think about the benefits of aviation within APAC, aviation will support 24.2 million jobs and contribute $516 billion to GDP by 906 commercial airports, 355 airlines, and 44 ANSPs.

To manage this growth in APAC, some $2 billion are being spent to improve ATM systems, infrastructure, equipage, and for training. Such cost is expected to increase safety, predictability, and capacity and to reduce aircraft time, noise, and complexity. However, we have to note that service improvement outcomes have not been defined so far and cross-border planning and improvement are not enough. He said that as ANSPs' collaboration is indispensable for airspace and whole route planning and cross-border solutions, IATA initiated APAC air traffic flow management project to achieve cross-border ATFM in 2014. The project succeeded in providing a baseline view of ATFM capability and interoperability in the regional cross borders.

Then he commented the status of present seamless ATM plan. Although the plan is agreed by all states concerned, implementation is very slow; generally each state is still thinking and planning within the own territories only. The reality is that we have a long way to achieve seamless sky.

He concluded that payment for ATM modernization must be used more efficiently under cooperative environment among nations concerned. And, in order to achieve sustainable economic performance, the aviation industry and governments must work together to lower costs and regulatory barriers.

3.7 Christian Schleifer-Heingärtner, "Standards to Ensure Global Harmonization and Worldwide Interoperability"

Christian Schleifer-Heingärtner is the secretary general of EUROCAE (European Organization for Civil Aviation Equipment). His speech focused on the importance of standards. First, he commented the close relationship between EUROCAE and ENRI. He said that ENRI paved the way from Europe to Asia as first EUROCAE Asia/Pacific member from 2011 and is contributing to EUROCAE working group activities.

He introduced the EUROCAE briefly. EUROCAE is a standardization body based on EU in which 175 members participate and 36 working groups are in active with 1400 experts. The EUROCAE members are from world manufactures, regulators, service providers, R&D organizations, international bodies as ICAO, and other standardization bodies as RTCA (Radio Technical Commission for Aeronautics). Its domains of activities spread almost all aviation systems as avionics, communication, navigation, surveillance, ATM systems, airports, etc. Technical work programs are organized to provide strategic vision of the activities in the context of current environment. About half of the total programs are carried out jointly with RTCA.

He spoke about ATM standardization strategy to modernize Global ATM Systems. World ATM modernization programs as NextGen, SESAR, CARATS, etc. are designed to be compatible with ICAO ASBUs but still have difference with each other because each program gives different priorities in its implementation and operation plans. He then presented next steps as SESAR2020 and GANP update in which standardization coordination is taken into account.

He spoke about smooth transfer from R&D to standardization and to deployment. Mutual understanding and close cooperation are indispensable in standardization and regulation processes among R&D organizations, standardization bodies, regulators, and service providers so that R&D results can be employed and deployed in smooth and harmonized manner. He emphasized that standards are the only way to avoid divergence between major R&D programs as NextGen, SESAR.

He presented EUROCAE's global coordination with other standardization bodies as ICAO, EASA (European Aviation Safety Agency), RTCA, etc. For example, EUROCAE standards published as EDs (EUROCAE Documents) are designed to show how to comply with regulations, acceptable means of compliance, best

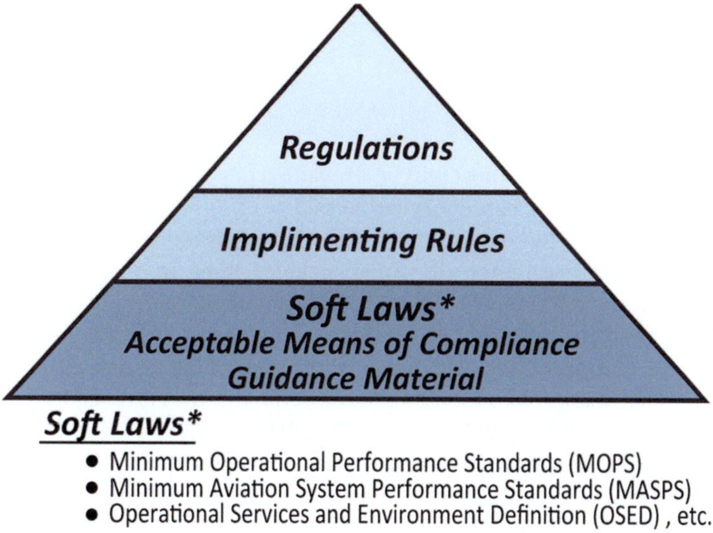

Fig. 2 Relationship between regulations, implementing rules, and soft laws

industry practice, and soft laws as MASPS (Minimum Aviation System Performance Specification), OSED (Operational Services and Environmental Definition), MOPS (Minimum Operational Performance Specification), etc. Figure 2 depicts the structure showing the relationship between regulations, implementing rules and soft laws. He added that ICAO, EASA, FAA, and other regulators recently refer to EDs and RTCA DOs (DOcuments) to construct ICAO's Annexes, EASA's ETSOs and FAA's TSOs, and so on. By that manner, we can avoid inconsistency in these regulations and standards.

In conclusion, he said that the EUROCAE is now recognized as an international standardization organization because performance-based standards and regulations have become more and more dominant in aviation. He stressed that the EUROCAE is calling for support of world experts to extend its activities.

3.8 Edward Bolton, "The NextGen Role in Global Research and Development Coordination"

Edward L. Bolton Jr. is the assistant administrator for NextGen at the Federal Aviation Administration (FAA), USA. His presentation focused on the update of NextGen. First, he spoke about NextGen's major objective of achieving successful transition from a present ground-based air traffic system to a smarter, satellite-based, digital-based, and performance-based system with improved information sharing to all users including controllers, operators, and flight crews.

He showed estimated benefits of NextGen capabilities from 2013 to 2030 as:

- Direct airline benefits: $51.4 billion
- Industry benefits: $2.4 billion
- Social benefits: $80.1 billion

Among the social benefits, most will be forwarded to passengers ($79.7 billion) by flight time saving and increased service, he said.

He then described the organization of the NextGen office. The office consists of six operating offices, each of which includes three to six divisions. He is responsible for integrating new and existing technologies and procedures into the National Airspace System (NAS) to reduce delay, save fuel, and lower aircraft emission. He introduced the FAA technical center which is under the NextGen organization. Its major role is to support advancement of the NextGen by integrating developed systems and capabilities through collaboration with industry, academia, and government and by providing the gateway for NAS upgrades and improvement.

Recently, the FAA, in collaboration with its stakeholders as airlines, pilots, air traffic controllers, and manufactures, produced a list of NextGen Priorities (the NextGen Joint Implementation Plan). Under the plan, the FAA and the aviation industry share responsibility to meet specific milestones and metrics for "high benefit, high readiness" NextGen initiatives. The initiatives include multiple runway operations, performance-based navigation, and surface and data communications. They have indicated a new approach for how to work together with industry.

He presented some new drivers as: commercial space, unmanned aircraft systems, cybersecurity, international collaboration, interagency coordination, equipage, and fiscal uncertainty. These drivers will be added in the process of NextGen modernization to analyze the entire system. He added that the modernization must be conducted in conjunction with operational and procedural changes. As for NextGen international collaboration, he cited several examples as:

- Future Air Traffic System development with Japan
- Supporting ATM Center of Excellence with Singapore
- SESAR-NextGen Coordination Committee with EU
- Mini Global Demonstrations of information exchange with Asia Pacific partners

In conclusion, he announced the establishment of a new NextGen International office. The office establishes NextGen International goals, aligns resources and efforts with goals, and reviews external initiatives and opportunities. He said that historically, NextGen international has been watching Europe. However, the office has recognized the importance of Asian nations. Thus, collaboration with NextGen will be easier and more familiar for Asia Pacific organizations with similar objectives.

3.9 *Akbar Sultan, "Collaborative R&D for NextGen and Global Harmonization to Inform ASBUs Blocks 2 and 3"*

Akbar Sultan is the deputy director of the Airspace Operations and Safety Program at NASA's Aeronautics Research Mission Directorate. His speech was to respond to Mr. Edward Bolton's address about NextGen Priorities and showed NASA Aeronautics' current six strategic research and technology programs:

1. Safe, efficient growth in global operations
2. Innovation in commercial supersonic aircraft
3. Ultraefficient commercial transports
4. Transition to low-carbon propulsion
5. Real-time system-wide safety assurance
6. Assured autonomy for aviation transformation

He then presented NASA's recent achievements relating to the above programs as:

1. Efficient Decent Adviser (EDA): EDA enables optimal profile descents at congested airports while metering, which brings about less fuel burn in approach phases in conjunction with continuous descent arrival (CDA).
2. ATM Technology Demonstration (ATD-1): ATD-1 enables PBN procedures while metering through the integration of ADS-B enabled flight deck merging/spacing, terminal area precision scheduling, and controller-managed spacing.
3. Dynamic Weather Routing: It can deliver bad weather avoidable and time- saving trajectories for better weather routes to aircraft in flight.
4. Spot and Runway Departure Advisor: The advisor enables nonstop taxiing to departure runway and runway queue management.
5. In-Trail Procedures (ITP): ITP enables aircraft to achieve ADS-B-based efficient altitude and speed settings by flight deck in-trail climb and descent in the oceanic routes.
6. Precision Departure Release Capability (PDRC): PDRC stands for integrating aircraft off time predictions into traffic management advisor so that departing aircraft can fit in an overhead stream.

He mentioned that above achievements were transferred to stakeholders as FAA, airport authorities, airlines, etc. and were contributing to fuel efficiency improvement, increased traffic throughput, reduced delay, reduced noise, etc.

Apart from NASA Aeronautics' practical R&D activities and achievements, he announced the establishment of "International Forum for Aviation Research (IFAR)" with the purpose of developing a globally harmonized research concept for integrated arrival/departure/surface operations. IFAR provides the work plans with different technical maturity from Task 1 (information exchange by face-to-face or remote WebEx meetings) to Task 8 (disseminating R&D results to inform ICAO and users).

Finally, he emphasized the importance of the IFAR activities and asked us to encourage specialists in the aviation community worldwide to join the activities.

3.10 Kazuo Yamamoto, "ENRI's R&D Activities for Globally Harmonized ATM Environment"

Kazuo Yamamoto was the president of the Electronic Navigation Research Institute (ENRI), organizer of the EIWAC. His presentation was designed to present major R&D topics and activities in ENRI for improving air traffic systems in Japan and for harmonizing different standards, R&Ds, and implementations in the world.

He described the roles of ENRI first. ENRI is a national institute conducting R&D and test for future aviation systems. Contribution to building global, regional, and national technical standards through its R&D and test results is also a major role. Another important one is to support improvement of present aviation systems in Japan. ENRI is now involved in CARATS, a National Air Navigation Plan managed by JCAB to promote modernization of national air traffic systems, and provides updated information about R&D, standardization, and implementation. ENRI is also supporting JCAB in ICAO as an adviser.

He presented ENRI's major R&D activities in terms of: (1) airport operation, (2) globally interoperable systems and data, (3) optimum capacity and flexible flights, and (4) efficient flight path. Some typical results are described below:

1. GBAS Prototype: ENRI developed CAT-1 and CAT-3 GBAS prototypes and installed at Osaka Kansai and Ishigaki Airport respectively. The results showed that these prototypes satisfied performance requirements even in adverse ionospheric environment. ENRI is now participating in the standardization activities in ICAO.
2. Aeronautical Mobile Airport Communication System (AeroMACS): ENRI developed a prototype of Air/Ground data link based on WiMAX technology and installed at Sendai Airport for test. The results demonstrated that one hundred times faster data link was attained than the current VHF Digital Link (VDL). Standardization work is now in operation by ENRI, WiMAX Forum, and ICAO.
3. Adaptive flight simulation for User Preferred Route (UPR): ENRI presented a new flight simulator that calculates optimum route, avoiding conflict adaptively when UPR is employed. After performance evaluation of the simulator, Informal Pacific ATC Coordinating Group (IPACG) between USA and Japan agreed that the simulator can provide an interoperable flight procedure.
4. Arrival procedures based on PBN and GBAS Landing System (GLS): ENRI proposed new arrival procedures based on RNP/ILS and RNP/GLS for precise curved approach. The procedures were evaluated by Boeing 787 flight simulators. The results demonstrated the feasibility of the navigation data base for RNP/GLS, which was developed through the close cooperation between ENRI and operators.

In addition to the R&D and test results, practical harmonization activities in ENRI were presented. He showed worldwide research/training collaborations with French, German, US, Korean, and Thailand universities and laboratories. He said that ENRI is now enhancing cooperation with Asia Pacific (APAC) organizations because regional ATM modernization should be prioritized for seamless sky in APAC.

Finally, he stressed the importance of harmonization, which is the theme of EIWAC2015. He recommended looking into the differences in world aviation systems, operational procedures, airspaces, and standards and trying to reduce such differences.

4 Invited Talks for Global Harmonization

In the invited session, the specialists from France, Singapore, and Japan spoke about global harmonization. This section summarizes those speeches. The presentation slides are available at ENRI website [4].

4.1 Patrick SOUCHU, "Moving from SESAR R&D Activities to Implementation: An ANSP Perspective"

Patrick Souchu is the program director of DSNA (Direction des Services de la Navigation Aérienne) and a member of EUROCAE Council. In his speech, he focused on the harmonization between SESAR R&D activities and implementation.

He spoke about DSNA first. DSNA is the French Air Navigation Service Provider taking responsibility of French air traffic system by five area control centers, 36 terminal maneuvering areas (TMAs), and 73 control towers. DSNA has more than 7500 employees including over 3900 air traffic controllers for ensuring safety, improving efficiency, punctuality, and customer satisfaction. He explained that DSNA controlled 2,845,000 flights in 2014 and 37 %, 15 %, and 48 % of which was international, domestic, and overflights, respectively.

He described the roles DSNA has been playing in the SESAR. DSNA is now involved in 64 SESAR projects including 53 R&Ds, 8 large-scale demonstrations, two remotely piloted aircraft system (RPAS) projects, and one CYBER Security project with about 100 DSNA employees, including 60 operational experts. Additional 30 experts will also participate in the projects when SESAR trials are conducted.

ANSP alliance was founded by NATS (UK), DFS (Germany), ENAV (Italy), DSNA, etc. to support and promote SESAR projects. The alliance was recently enforced by adding members as ANS CR (Czech Republic), PANSA (Poland), skyguide (Switzerland), etc. He said that validation of SESAR R&D results is mainly driven by the alliance through exercises and demonstrations. He showed typical examples of validation activities as:

- Validation of flight object and ground-ground interoperability through DSNA-ENAV-DFS-MUAC (Maastricht Upper Area Control Centre) simulation
- Validation of Extended AMAN (E-AMAN) from central Europe to London-Heathrow based on EUROCAE standardization of initial SWIM for E-AMAN

He moved onto the topic of "ANSP involvement in SESAR deployment." According to European ATM Master Plan, many different actions have been taken during R&D and implementation as large-scale demonstration, pre-implementation, standardization, safety analysis, etc. SESAR Deployment European framework supports the implementation of the Master Plan through government mechanism, deployment program, common projects, implementation projects, and targeted incentives.

He explained the Pilot Common Project (PCP) which is designed to evaluate SESAR R&D results from the viewpoints of linkage to the ATM Master Plan, ATM functionalities, positive business case contribution, standardization and regulatory aspects, etc. Among them, ATM functionalities are assessed in terms of six subjects as:

1. Extended AMAN and PBN in high-density TMAs
2. Airport Integration and Throughput Functionalities
3. Flexible Airspace Management and Free Route
4. Network Collaborative Management
5. Ground-ground integration and aeronautical data management and sharing
6. Initial Trajectory Information sharing for air-ground integration toward initial four-dimensional (i4D) operation

Finally, he said that SESAR deployment is on successful steps under its concrete framework.

4.2 Mohamed Faisal Bin Mohamed SALLEH, "ATM Research and Development for Asia Pacific"

Mohamed Faisal Bin Mohamed Salleh is the deputy director, Air Traffic Management Research Institute (ATMRI) of Nanyang Technological University, Singapore. He was invited to EIWAC2015 as representative of ATMRI to address the importance of ATM research and development common in the Asia Pacific (APAC) region.

He first showed the video introducing ATMRI and then presented R&D topics that ATMRI is now involved in. ATMRI was established in 2013 in cooperation with Civil Aviation Authority of Singapore and Nanyang Technological University to promote ATM transformation and harmonization in the region by providing innovative solutions.

He said that APAC should be a community of common interest just like EU. And his APAC ATM outlook indicated the importance of R&D activities suited for

APAC because this region has distinctive cultural and political backgrounds and a geographical feature that large parts of the region are covered by ocean. He also pointed out the necessity of implementation of the seamless APAC ATM plan.

He presented the key challenges in APAC as:

- Lack of regional coordination by different interests and requirements of each state
- Route structure that has not kept pace with present traffic growth
- Insufficient communication, navigation, and surveillance capability
- Adverse weather in tropical region

To respond to such challenges, he proposed several examples of R&D opportunities as:

- Space-based ADS-B, full CPDLC, SWIM environment
- Enhanced operational decision through integrated meteorological information
- PBN (performance-based navigation) and RNP (required navigation performance)
- Dynamic airspace management with modernized route structure
- Integrated virtual ATFM

He informed ATMRI's recent establishment of ASEAN ATM simulation and modeling function that enables detailed analysis and prediction of future traffic arising from the implementation of ASEAN Single Aviation Market. He said that by 2033, the ASEAN sky may see traffic volumes similar to Europe operating 30,000 flights a day. Therefore, dependable future traffic prediction will be very important to restructure air routes in the ASESAN sky. This ATM simulation function also enables radar coverage prediction in South Chana Sea, so that we can improve the coverage by adding ADS-B capabilities in the region.

As a regional effort to realize global harmonization for future sky, he proposed to establish ASEAN ATM Master Plan based on ICAO Global Air Navigation Plan, APAC Seamless ATM Plan, and ASEAN National Master Plans. This Master Plan presents 9 focus areas as ATS route network collaborative management, civil/military cooperation, surveillance infrastructure, airport ATM operations, etc.

He concluded that communication and collaboration among all stakeholders is critical for a seamless ASEAN sky.

4.3 *Yasuhiro Koshioka, "DREAMS Project, Its Results, and Implementation Status"*

Yasuhiro Koshioka is director of Technology Demonstration Research Unit, Aeronautical Technology Directorate of Japan Aerospace Exploration Agency (JAXA). His address was about JAXA's DREAMS (Distributed and Revolutionary Efficient Air-Traffic Management System) Project that aims to promote ICAO ATM Operational Concept by providing its R&D results.

He first described the outline of JAXA. JAXA is a national research institute taking responsibility of research and development about space science/technology and aeronautical technology in Japan. Its financial support comes mainly from the Ministry of Education, Culture, Sports, Science, and Technology.

The DREAMS Project was organized from the background that Asian sky is now facing:

- The growth rate of Asian air traffic is higher than any other regions in the world, so we have to respond to such growth.
- In Japan, traffic concentration on metropolitan terminal areas and airports becomes heavier, so we must mitigate such congestion.

In the DREAMS Project, JAXA focused on developing key technologies to improve ATM in terminal areas. He presented the specific R&D subjects, key topics, and major results in the DREAMS Project as bellow:

1. Weather information (wake vortex and wind shear) technologies:
    - Wake vortex and wind shear detection systems and estimation of their hazards
    - Reduction of wake vortex separations, optimization of takeoff, and landing sequences
    - Development of LOTAS (Low-Level Turbulence Advisory System) and ALWIN (Airport Low-Level Wind Information) that can provide approaching aircraft with wind shear information
2. Noise abatement technologies:
    - Improvement of noise prediction through air to ground sound propagation analysis
    - Better noise prediction regardless of weather conditions and seasons, which has enabled approach path optimization
3. High-accuracy satellite navigation technologies:
    - GPS/INS integrated navigation technique and its performance test
    - GBAS availability improvement
4. Trajectory control technologies:
    - Evaluation of GBAS TAP (terminal area path) for precision curved approach
    - Autopilot and flight guidance procedures for GBAS TAP operation
5. Disaster-Relief Small Aircraft operation:
    - Establishment of operation procedures and information sharing in damaged area for disaster-relief small aircraft
    - Development of "D-NET" which enables information sharing and mission assignment among small aircraft and helicopters in disaster relief activities

He emphasized that most of above results were not only implemented in airports or in operations but provided in the ICAO working groups, RTCA special committees for standardization. He also presented the flow from research, development, demonstration, and evaluation and then to implementation defined in the DREAMS Projects.

He concluded that smooth transfer from R&D results to implementation can be achieved only by constant information sharing and collaboration among all parties concerned.

5 Conclusions

The fourth ENRI International Workshop on ATM and CNS (EIWAC 2015) was held in November 2015 with a view to sharing comprehensive information on the latest ATM/CNS technologies and operations among the participants and seeking potential partners for R&D, standardization, and global harmonization activities.

History and overview of the EIWAC series were described first. Then the topics and opinions presented by the keynote and invited speakers who belong to various organizations as regulators, standardization bodies, ANSPs, operators, and R&D institutes were summarized. The keynote and invited speeches have shown that the speakers have common recognition about the problems that present aviation systems are facing, but have diverse views for solving the problems and for modernizing the aviation systems. Figure 3 is a photo showing the EIWAC 2015 speakers and organizers.

Fig. 3 EIWAC 2015 speakers and organizers

Finally, we would like to announce the recent restructuring that ENRI was undertaken. In April 2016, ENRI was merged with two other national research institutes belonging to MLIT and became a department of the National Institute of Maritime, Port and Aviation Technology (MPAT). However, the department had been named "ENRI" and succeeded to the total role that former ENRI was in charge. Furthermore, "new ENRI" is trying to take this opportunity to expand its areas of activities.

ENRI is now planning to hold next EIWAC in 2017. The next one will also attract many people in the aviation community worldwide and can contribute to the advancement of global aviation systems.

References

1. EIWAC2009 homepage, http://www.enri.go.jp/eiwac/2009/en/
2. EIWAC2010 homepage, http://www.enri.go.jp/eiwac/2010/en_EIWAC2010siryou.html
3. EIWAC2013 homepage, http://www.enri.go.jp/eiwac/en_EIWAC2013siryou.html
4. EIWAC2015 homepage, http://www.enri.go.jp/eiwac/eiwac_2015_eng.html
5. Electronic Navigation Research Institute, Ed. (2014) Air Traffic Management and Systems, Springer: 3–14.
6. International Civil Aviation Organization (2014) 2013–2028 Global Air navigation Capacity & Efficiency Plan, 2014–2016 Triennium Edition DOC 9750: 10–11.

Part II
Trajectory Planning and Optimization

Large-Scale 4D Trajectory Planning

Arianit Islami, Supatcha Chaimatanan, and Daniel Delahaye

Abstract To sustain the continuously increasing air traffic demand, the future air traffic management system will rely on a so-called trajectory-based operations concept that will increase air traffic capacity by reducing the controller workload. This will be achieved by transferring tactical conflict detection and resolution tasks to the strategic planning phase. In this future air traffic management paradigm context, this paper presents a methodology to address such trajectory planning at nationwide and continent scale. The proposed methodology aims at minimizing the global interaction between aircraft trajectories by allocating alternative departure times, alternative horizontal flight paths, and alternative flight levels to the trajectories involved in the interaction. To improve robustness of the strategic trajectory planning, uncertainty of aircraft position and aircraft arrival time to any given position on the trajectory are considered. This paper presents a mathematical formulation of this strategic trajectory planning problem leading to a mixed-integer optimization problem, whose objective function relies on the new concept of interaction between trajectories. A computationally efficient algorithm to compute interaction between trajectories for large-scale applications is presented and implemented. Resolution method based on hybrid-metaheuristic algorithm has been developed to solve the above large-scale optimization problem. Finally, the overall methodology is implemented and tested with real air traffic data taking into account uncertainty over the French and the European airspace, involving more than 30,000 trajectories. Conflict-free and robust 4D trajectory planning is produced within computational time acceptable for the operation context, which shows the viability of the approach.

Keywords Aircraft 4D trajectory • Strategic planning • Hybrid metaheuristic

A. Islami • S. Chaimatanan (✉) • D. Delahaye
Applied Mathematics Laboratory, École Nationale de l'Aviation Civile (ENAC), 7 Avenue Edourd Belin, 31055, Toulouse, France
e-mail: arianit.islami@caa-ks.org; supatcha@gistda.or.th; delahaye@recherche.enac.fr

Electronic Navigation Research Institute, *Air Traffic Management and Systems II*, Lecture Notes in Electrical Engineering 420, DOI 10.1007/978-4-431-56423-2_2

1 Introduction

Air traffic regulations impose that aircraft must always be separated by some prescribed distance, noted N_v for the vertical separation and N_h for the horizontal separation. Aircraft are considered to be in *conflict* when these *minimum separation* requirements are violated. As the global air traffic demand keeps on increasing, congestion problem becomes more and more critical. One of the key solutions is to balance the air traffic demand and the overall capacity of the Air Traffic Management (ATM) system. In order to cope with the increasing demand, the future ATM system will rely on the trajectory-based operations concept. In this concept, aircraft will be required to follow a negotiated *conflict-free* trajectory, accurately defined in four dimensions (three spatial dimensions and time) in order to reduce the need of controller's intervention during the tactical phase. In this perspective, the key factor to improve the ATM capacity is an efficient strategic 4D trajectory planning methodology to compute a *conflict-free 4D trajectory* for each aircraft.

In this work, we propose a methodology to address such a strategic planning of trajectories at national and continent scale. The goal of the proposed method is to separate a given set of aircraft trajectories in both the three-dimensional space and in the time domain by allocating an alternative flight plan (route, departure time, and flight level) to each flight.

Instead of trying to satisfy the capacity constraint, we focus on minimizing the global *interaction* between trajectories. An interaction between trajectories occurs when two or more trajectories have an effect on each other, for instance, when trajectories occupy the same space at the same period of time. Therefore, contrary to the concept of conflict, the measurement of interaction does not only refer to the violation of minimum separation requirements. It also allows us to take into account other separation criteria such as minimum separation time between aircraft crossing at the same point.

In real-life situations, aircraft may not be able to follow precisely the assigned 4D trajectory due to external events, such as passenger delays, wind conditions, etc. Besides, aircraft may not be able to fly at their optimal speed profile in order to satisfy the hard constraints imposed on the 4D trajectory. To improve robustness of the deconflicted trajectories and to relax the 4D trajectory constraints, uncertainties of aircraft position and arrival time will also be taken into account in the strategic trajectory planning process presented in this paper.

The following section of this paper is organized as follows. Section 2 reviews previous related works on aircraft trajectory deconfliction. Section 3 describes uncertainty model, explains the concept of interaction between trajectories, and presents the trajectory planning problem in mathematical framework. Section 4 proposes an efficient method for detecting interactions between aircraft trajectories in a large-scale context. Section 5 presents a hybrid-metaheuristic optimization algorithm, which relies on simulated annealing and on a hill-climbing local-search method, to solve the problem. Finally, numerical results are presented and discussed in Sect. 6.

2 Previous Related Works

During recent years, there are many research works in the literature that address the trajectory deconfliction problem considering large-scale air traffic. Aircraft trajectory deconfliction problem that relies on genetic algorithm (GA) to solve en route conflicts between trajectories, taking into account uncertainties of aircraft velocity, is considered in [12]. The authors propose two conflict-resolution maneuvers: modifying the heading and modifying the flight level. The solutions are provided by GA. It is able to solve all conflicts involving 7,540 flights considering different levels of uncertainties within reasonable computation time.

In [2], the authors consider a 4D trajectory deconfliction problem using a ground-holding method. Potential conflicts between trajectories are detected by pairwise comparison. However, in the presence of takeoff time uncertainties, the proposed method must allocate significant delays in order to solve all the conflicts. To increase the degrees of freedom, the same authors introduce an option to allocate alternative flight levels in [3]. The results show advantages of using the flight level allocation technique in terms of reduced delay, in the presence of departure time uncertainties. In [1], a flight-level allocation technique is used to address 4D trajectory deconfliction at the European continent scale. However, the proposed method yields residual potential conflicts.

Another idea to separate trajectories is based on speed regulations; it is used, for instance, in [8]. In these works, conflict detection and resolution are performed at two layers with different sampling periods and time windows. Speed regulations introduce additional degree of freedom to the trajectory design. However, it requires numerous extensive and fine-tuned computations, which are not suitable to implement in a large-scale problem.

In [9, 10], a light propagation algorithm (LPA) is introduced to solve potential conflicts between 4D trajectories and to avoid congested and bad-weather areas. The optimal trajectories that solve conflicts are provided by a branch and bound (B&B) algorithm.

In [10], to improve robustness of aircraft trajectories, uncertainty is modeled as a time segment. The uncertainty increases the difficulty of the problem and reduces the solution space, so that the LPA can remove only 88% of the conflicts. The remaining conflicts are solved by imposing time constraints called required time of arrival (RTA).

A methodology to optimize and deconflict aircraft trajectories in the horizontal plane, in en route environment, and in real time is proposed by the author of [13]. In this work, aircraft trajectories are deconflicted and optimized in the time scale of thirty minutes into the future by sequentially computing optimal-wind and conflict-free trajectories for each aircraft, considering previously planned trajectories as obstacles.

However, none of the proposed methodologies is able to solve globally the trajectory deconfliction problem due to its size and complexity. Most of the algorithms proposed in the literature rely on the moving time window strategy to reduce the size of the problem. This strategy is effective for conflict detection and resolution in tactical phases. However, when high-density traffic is involved, it tends to fail to solve all conflicts.

In this work, we put forward the works presented in [4–7]. The proposed 4D trajectory planning methodology aims at solving conflict between all involving trajectories simultaneously at strategic level. In these works, optimal 4D trajectories for individual flights were allocated by solving a combinatorial optimization problem using a non-population-based hybrid-metaheuristic optimization method.

3 Mathematical Modeling

This section presents the mathematical model used to describe our strategic trajectory planning methodology. First, uncertainty of aircraft positions and arrival times based on two different models are characterized. Then, a definition of *interaction* between trajectories is given. Next, the route/departure time/flight level allocation techniques adapted for strategic trajectory planning are described. Finally, a mathematical formulation of the interaction minimization problem is presented.

3.1 Uncertainties

Consider a given set of N trajectories, where each trajectory, i, is defined by a time sequence of 4D coordinates, $P_{i,k}(x_{i,k}, y_{i,k}, z_{i,k}, t_{i,k})$, for $k = 1, \ldots, K_i$, where K_i is the total number of *sampling points* of trajectory i, for $i = 1, \ldots, N$. Each trajectory is sampled with a (given) constant sampling time, Δt. These coordinates specify that aircraft i must arrive at a given point $(x_{i,k}, y_{i,k}, z_{i,k})$ at time $t_{i,k}$.

However, in reality, aircraft are subjected to unpredicted external events, which cause uncertainties on aircraft position and arrival time with respect to their planned 4D trajectory. In order to consider such uncertainties, we rely on the concept of robust optimization, using two different models of the uncertainty sets.

3.1.1 Deterministic Model

For simplicity, we define x_P, y_P, z_P, t_P as the 4D coordinate of any given point $P_{i,k}$. Consider an initial 4D trajectory planning specifying that an aircraft must arrive at a given horizontal coordinate (x_P, y_P) at time t_P. Due to uncertainties, we shall assume that the *real* horizontal position, (x_P^r, y_P^r), of the aircraft at time t_P can be in an area defined by a disk of radius R_h (defined by the user) around (x_P, y_P), as illustrated in Fig. 1. In other words, the possible locations of the aircraft at time t_P are the elements of the set: $\{(x_P^r, y_P^r) : (x_P^r - x_P)^2 + (y_P^r - y_P)^2 \leq R_h^2\}$.

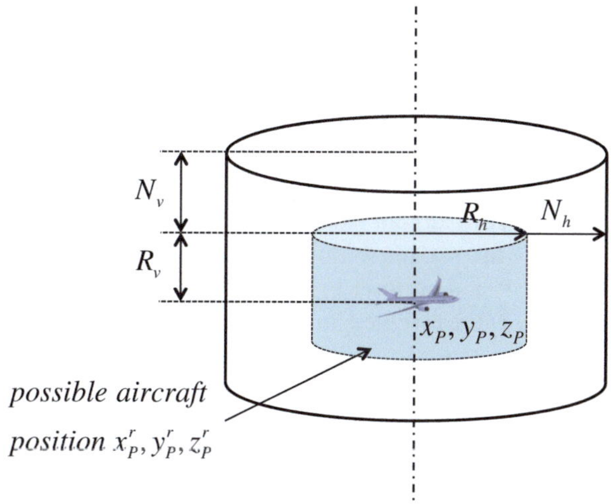

Fig. 1 Possible aircraft position in the 3D space domain in the presence of deterministic uncertainty

To ensure horizontal separation of aircraft subjected to such uncertainties, the protection volume has to be enlarged by a radius of R_h as illustrated in Fig. 1. Thus, the *robust minimum separation in the horizontal plane*, N^r_h, is defined as: $N^r_h := N_h + R_h$, where N_h is the minimum horizontal separation of the case without uncertainty.

In the vertical dimension, we shall assume that during such a non-level flight phase, the real altitude, denoted z^r_P, of the aircraft at a given time t_P lies in a bounded interval defined by an uncertainty radius R_v (set by the user) which reduces strongly when the aircraft reaches its requested flight level. In other words, the possible altitudes of the aircraft during non-level flight phase at time t_P are the elements of the set: $\{z^r_P : z_P - R_v \leq z^r_P \leq z_P + R_v\}$.

To ensure vertical separation of aircraft subjected to such uncertainties, the vertical separation requirement has to be enlarged by R_v as illustrated in Fig. 1. Thus, the *robust minimum separation in the vertical dimension*, noted N^r_v, is defined as: $N^r_v := N_v + R_v$, where N_v is the minimum vertical separation of the case without uncertainty.

In addition to the uncertainty in the 3D space domain (see Fig. 1), aircraft may be subjected to uncertainty so that it arrives at a given position with a time error. Let t_ϵ be the *maximum time error* (defined by the user). For simplicity, to implement the interaction detection scheme, we shall assume that t_ϵ is chosen so that it is a multiple of the discretization time step Δt. The real arrival time, noted t^r_P, of aircraft at the same trajectory point therefore lies in the time interval: $[t_P - t_\epsilon, t_P + t_\epsilon]$, where t_P is the assigned arrival time to point $P_{i,k}$.

3.1.2 Probabilistic Model

The worst-case-oriented uncertainty model presented above considers that every possible cases in the given uncertainty set are equally likely. However, some events corresponding to the points in the uncertainty set have very low probability to occur. Trying to immune the solution against such events could yield unnecessarily costly solutions and can be interpreted as too conservative for a situation involving high levels of uncertainty as it is the case in strategic planning.

An aircraft is able to follow a given flight profile with very high accuracy thanks to the flight management system (FMS).We shall consider that the residual uncertainty of aircraft position is more likely to occur in the time domain.

Using the maximum time error, t_ϵ (set by the user), the *predicted* arrival time of an aircraft at a position $P_{i,k}$ under uncertainty lies in the interval: $[t_P - t_\epsilon, t_P + t_\epsilon]$, where t_P is the assigned arrival time to point $P_{i,k}$.

For the purpose of interaction computation, which will be explained in the following subsection, we assume here that the predicted aircraft arrival time can be modeled as a random variable with the following triangular distribution defined over the interval $[t_P - t_\epsilon, t_P + t_\epsilon]$. Given the lower limit $t_P - t_\epsilon$, the upper limit $t_P + t_\epsilon$, the *predicted* arrival time, to the position $P_{i,k}$ is given by the probability density function:

$$\mathcal{T}_{t_P,t_\epsilon}(t) = \begin{cases} 0 & \text{for } t < t_P - t_\epsilon, \\ \frac{(t - t_P + t_\epsilon)}{t_\epsilon^2} & \text{for } t_P - t_\epsilon \leq t \leq t_P, \\ \frac{(-t_P + t_\epsilon - t)}{t_\epsilon^2} & \text{for } t_P < t \leq t_P + t_\epsilon, \\ 0 & \text{for } t_P + t_\epsilon < t, \end{cases} \tag{1}$$

where $\mathcal{T}_{t_P,t_\epsilon}(t)$ denotes the triangular distribution. Figure 2 illustrates the uncertainty of arrival time of two aircraft A and B to the trajectory sample points P and Q, respectively. The time uncertainties are defined by a triangular distribution function over the time interval $[t_P - t_\epsilon, t_P + t_\epsilon]$ and $[t_Q - t_\epsilon, t_Q + t_\epsilon]$, respectively.

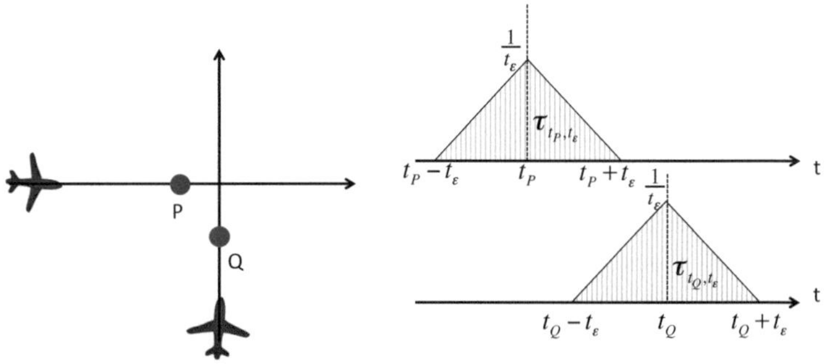

Fig. 2 Uncertainty of aircraft arrival time, defined by triangular distribution over given time intervals (*left*: view in the space domain; *right*: view in the time domain)

3.2 Interaction Between Trajectories

The concept of interaction between trajectories is introduced in [7]. It is a measurement that indicates when two or more trajectories occupy the same space at the same period of time. It is different from the *conflict* situation, which corresponds simply to a violation of the minimum *separation* (i.e., 5 NM horizontally and 1,000 ft. vertically). Additional separation conditions, such as time separation, topology of trajectory intersection, distance between trajectories, etc., can also be taken into account in the concept of interaction.

To explain the process to determine the interaction between aircraft trajectories, let us first consider two trajectories A and B, and let P and Q be any pair of sample points on the trajectories A and B, respectively. To consider the abovementioned deterministic uncertainty models, we must check whether the minimum separations, N_h^r and N_v^r, are satisfied, between every possible pair of points such as P and Q (pairwise comparisons). A potential conflict between trajectories A and B, taking into account uncertainties, can occur when the three following conditions are satisfied for a certain pair of sample points, P and Q, from each trajectory:

- $d_h := \sqrt{(x_P - x_Q)^2 + (y_P - y_Q)^2} < N_h^r$.
- $d_v := |z_P - z_Q| < N_v^r$.
- $[t_P - t_\epsilon, t_P + t_\epsilon] \cap [t_Q - t_\epsilon, t_Q + t_\epsilon] \neq \emptyset$, i.e. $|t_P - t_Q| \leq 2t_\epsilon$.

When the above conditions are satisfied, we say that *point P is in conflict with point Q taking into account the deterministic-type uncertainty.*

Let us define further

$$C^D(P,Q) = \begin{cases} 1 & \text{if point } P \text{ is in conflict with point } Q \\ 0 & \text{otherwise.} \end{cases} \tag{2}$$

With the above definitions, the *interactions at point* $P_{i,k}$, denoted $\Phi_{i,k}^D$, may be defined as the total number of times the protection volume around point $P_{i,k}$ taking into account the deterministic-type uncertainty is violated. Therefore, $\Phi_{i,k}^D$ is given by

$$\Phi_{i,k}^D = \sum_{\substack{j=1 \\ j\neq i}}^{N} \sum_{l=1}^{K_j} C^D(P_{i,k}, P_{j,l}), \tag{3}$$

where K_i is the number of sampled points of trajectory i.

Finally, the *robust total interaction between trajectories*, that we are minimizing, is

$$\Phi_{tot}^D = \sum_{i=1}^{N} \sum_{k=1}^{K_i} \Phi_{i,k}^D, \tag{4}$$

where N is the total number of trajectories.

To explain the process to compute the total robust interaction between trajectories based on probabilistic-type uncertainty, let us consider the trajectories A and B given in Fig. 2. Let P and Q be any trajectory sample points on trajectories A and B, respectively. The predicted arrival time of aircraft A to the given point P and the predicted arrival time of aircraft B to the given point Q are given by $\mathcal{T}_{t_P,t_\epsilon}(t)$ and $\mathcal{T}_{t_Q,t_\epsilon}(t)$, respectively.

Again, a potential conflict between trajectories A and B occurs when there exists a pair of points, P and Q, from each trajectory such that the three following conditions are satisfied:

- $d_h < N_h^r$;
- $d_v < N_v^r$;
- and $[t_P - t_\epsilon, t_P + t_\epsilon] \cap [t_Q - t_\epsilon, t_Q + t_\epsilon] \neq \emptyset$.

The probabilistic interaction, denoted $\mathcal{P}_{t_\epsilon}(P, Q)$, associated to the trajectory sample points P and Q is formally defined as follows:

$$\mathcal{P}_{t_\epsilon}(P, Q) := \int_{I_{PQt_\epsilon}} \mathcal{T}_{t_P,t_\epsilon}(t)\mathcal{T}_{t_Q,t_\epsilon}(t)dt, \tag{5}$$

where I_{PQt_ϵ} denotes the time interval $[t_P - t_\epsilon, t_P + t_\epsilon] \cap [t_Q - t_\epsilon, t_Q + t_\epsilon]$. Remark that when this intersection is the empty set, the integral in (5) is reduced to zero.

With the above definition, we define a *robust interaction at a point* $P_{i,k}$ based on the probabilistic-type uncertainty, denoted $\Phi_{i,k}^P$, to be the sum of all the probabilistic interaction associated to point P.

Hence, we have

$$\Phi_{i,k}^P := \sum_{\substack{j=1 \\ j \neq i}}^{N} \sum_{l=1}^{K_j} \mathcal{P}_{t_\epsilon}(P_{i,k}, P_{j,l}), \tag{6}$$

where K_j is the number of sampling points for trajectory j and where $\mathcal{P}_{t_\epsilon}(P_A, P_B)$ is the probabilistic interaction associated to the sample points P_A and P_B of trajectory A and B, respectively.

Therefore, the *total interaction between trajectories*, based on probabilistic-type uncertainty, denoted Φ_{tot}^P, for a whole N-aircraft traffic situation is simply defined as

$$\Phi_{tot}^P = \sum_{i=1}^{N} \Phi_i^P = \sum_{i=1}^{N} \sum_{k=1}^{K_i} \Phi_{i,k}^P. \tag{7}$$

3.3 Route/Departure Time/Flight Level Allocation

The objective of this work is to allocate an alternative trajectory, an alternative departure time, and an alternative flight level for each aircraft in order to minimize the total interaction between trajectories, taking into account uncertainty of aircraft position and time.

Given data. A problem instance is given by:

- A set of initial N discretized 4D trajectories;
- The discretization time step, Δt
- The number of allowed virtual waypoints, M
- The maximum allowed advance departure time shift of each flight i, $\delta_a^i < 0$
- The departure time shift step size, δ_s
- The maximum allowed delay departure time shift of each flight i, $\delta_d^i > 0$
- The maximum allowed flight level shift of each flight i, $l_{i,max}$
- The maximum allowed route length extension coefficient of each flight i, $0 \leq d_i \leq 1$
- The length of the initial en route segment of each flight i, $L_{i,0}$

The alternative departure time, the alternative route, and the alternative flight level to be allocated to each flight are modeled as follows.

Alternative departure time. The departure time of each flight can be shifted by a positive (delay) or a negative (advance) time shift. Let $\delta_i \in \Delta_i$ be a departure time shift attributed to flight i, where Δ_i is a set of acceptable time shifts for flight i. The departure time t_i of flight i is therefore $t_i = t_{i,0} + \delta_i$, where $t_{i,0}$ is the initially planned departure time of flight i. The departure time shift δ_i will be limited to lie in the interval $\Delta_i := [\delta_a^i, \delta_d^i]$. Common practice in airports conducted us to rely on a discretization of this time interval using time shift step size δ_s. This yields $N_a^i := \frac{-\delta_a^i}{\delta_s}$ possible advance slots and $N_d^i := \frac{\delta_d^i}{\delta_s}$ possible delay slots of flight i. Therefore, we define the set, Δ_i, of all possible departure time shifts of flight i by

$$\begin{aligned}\Delta_i :=&\{-N_a^i.\delta_s, -(N_a^i - 1).\delta_s, \dots, \\ &- \delta_s, 0, \delta_s, \dots, (N_d^i - 1).\delta_s, N_d^i.\delta_s\}.\end{aligned} \tag{8}$$

Alternative trajectory design. In this work, an alternative trajectory is constructed by placing a set of virtual waypoints, denoted

$$w_i = \{w_i^m | w_i^m = (w_{ix'}^m, w_{iy'}^m)\}_{m=1}^M, \tag{9}$$

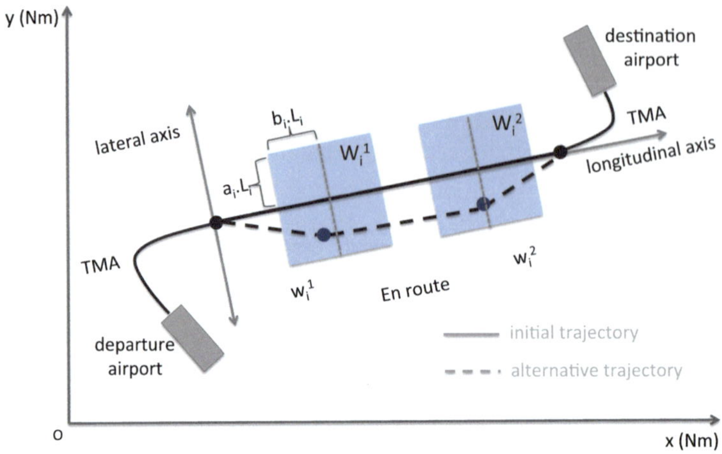

Fig. 3 Initial and alternative trajectories with rectangular shape possible location of $M = 2$ virtual waypoints

near the initial en route segment and then by reconnecting the successive waypoints with straight-line segments as illustrated in Fig. 3. To limit the route length extension, the alternative en route profile of flight i must satisfy

$$L_i(w_i) \leq (1 + d_i), \tag{10}$$

where $L_i(w_i)$ is the length of the alternative en route profile determined by w_i. Figure 3 illustrated initial and alternative trajectories, constructed with $M = 2$ waypoints, where the location of each waypoint is constrained to be in a rectangular shape possible location. Let $W_{ix'}^m$ be a set of all possible normalized longitudinal locations of the mth virtual waypoint on trajectory i. For each trajectory i, the normalized longitudinal component, $w_{ix'}^m$, is set to lie in the interval:

$$W_{ix'}^m := \left[\left(\frac{m}{1+M} - b_i \right), \left(\frac{m}{1+M} + b_i \right) \right], \tag{11}$$

where b_i is a (user-defined) parameter that defines the range of possible normalized longitudinal component of the mth virtual waypoint on trajectory i. To obtain a regular trajectory, the normalized longitudinal component of two adjacent waypoints must not overlap, i.e.,

$$\left(\frac{m}{1+M} + b_i \right) < \left(\frac{m+1}{1+M} - b_i \right) \tag{12}$$

and hence the user should choose b_i so that

$$b_i < \frac{1}{2(M+1)}. \tag{13}$$

Let $W^m_{iy'}$ be a set of all possible normalized lateral locations of the mth virtual waypoint on trajectory i. Similarly, the normalized lateral component, $w^m_{iy'}$, is restricted to lie in the interval:

$$W^m_{iy'} := [-a_i, a_i], \tag{14}$$

where $0 \leq a_i \leq 1$ is a (user-defined) model parameter that defines the range of possible normalize lateral location of the mth virtual waypoint on trajectory i, chosen a priori so as to satisfy (10). More detail about the method to modify the trajectory is presented in [7].

Alternative flight level. Another variable to modify the trajectory of each flight i is a flight level shift $l_i \in \mathbb{Z}$. Therefore, the flight level, FL_i, of flight i is given by: $FL_i = FL_{i,0} + l_i$, where $FL_{i,0}$ is the initially planned flight level of flight i. Figure 4 shows a trajectory with two alternative flight levels. In order to limit the change of flight levels, the set, ΔFL_i, of all possible flight level shifts for flight i is set to

$$\Delta FL_i := [FL_{i,0} - l_{i,max}, \ldots, 0, \ldots, FL_{i,0} + l_{i,max}], \tag{15}$$

where $l_{i,max}$ is the (user-provided) maximum flight level shifts allowed to be allocated to flight i.

Let us set the compact vector notation: $\boldsymbol{\delta} := (\delta_1, \delta_2, \ldots, \delta_N)$, $\mathbf{w} := (w_1, w_2, \ldots, w_N)$, and $\mathbf{l} := (l_1, l_2, \ldots, l_N)$. We shall denote by u_i the components

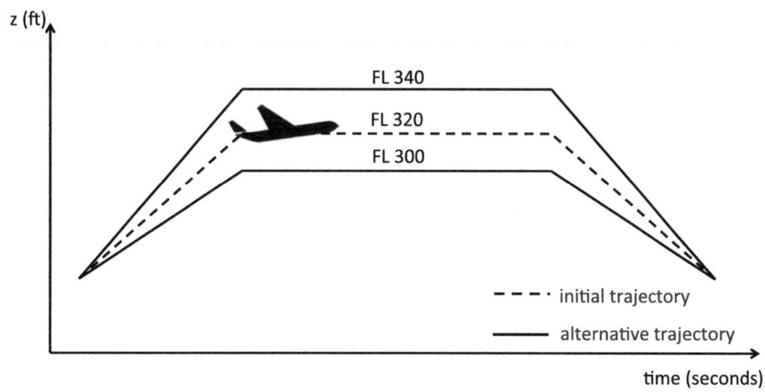

Fig. 4 Two alternative vertical profiles for a trajectory (two alternative flight levels)

of u. It is a vector whose components are related to the modification of the ith trajectory; therefore, our decision variable is

$$u := (\boldsymbol{\delta}, \mathbf{l}, \mathbf{w}).$$

The strategic trajectory planning problem under uncertainty can be represented by an interaction minimization problem formulated as a mixed-integer optimization problem as follows:

$$\begin{aligned}
&\min_u \Phi_{tot}(u) \\
&\text{subjectto} \\
&\quad \delta_i \in \Delta_i, \quad i = 1, 2, \ldots, N \\
&\quad l_i \in \Delta FL_i, \quad i = 1, 2, \ldots, N \\
&\quad w_i^m \in W_{ix'}^m \times W_{iy'}^m, \quad m = 1, 2, \ldots, M, \\
&\qquad\qquad\qquad\qquad\qquad i = 1, 2, \ldots, N,
\end{aligned} \tag{16}$$

where $\Phi_{tot}(u)$ is defined by (4) or (7) according to uncertainty model under consideration, and Δ_i, $W_{ix'}^m$, and $W_{iy'}^m$, ΔFL_i are defined by (8), (11), (14), and (15), respectively.

4 Interaction Detection

In order to evaluate the objective function, we rely on a grid-based interaction detection scheme which is implemented in a so-called *hash table* as presented in [4, 6, 7].

First, the airspace is discretized using a 4D grid (3D space + time), as illustrated in Fig. 5. The size of each cell in the x, y, z, and t direction is defined by the minimum separation requirement, N_h^r, N_v^r, and the discretization time step, Δt. To detect conflicts, the idea is to store the N trajectories in each corresponding cell in the 4D grid. Then, for each trajectory i, and for each cell (I_x, I_y, I_z, I_t) corresponding

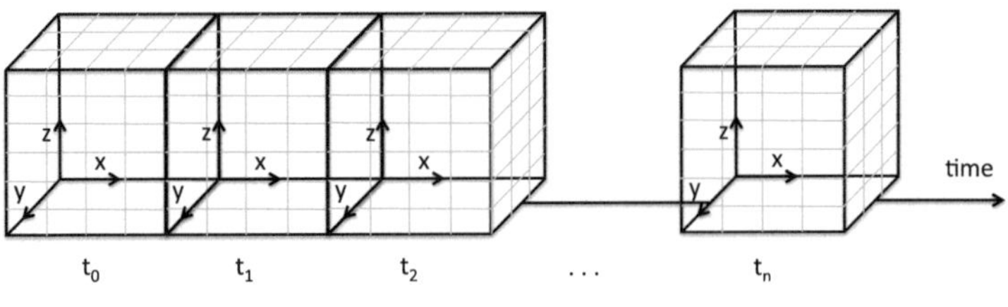

Fig. 5 Four-dimension (3D space - time) grid for conflict detection

to each sampling point $P_{i,k} := (x_P, y_P, z_P, t_P)$, we simply need to check all the surrounding (adjacent) cells in the x, y, and z directions corresponding to the time period $[t_p - 2t\epsilon, t_P + 2t\epsilon]$. If one of these surrounding cells is occupied by another aircraft, for instance j, we then note $j \in (I_x, I_y, I_z, I_t)$, and then the horizontal distance, d_h, and the vertical distance, d_v, between point $P_{i,k}$ and the sample point corresponding to aircraft j are computed.

A violation of protection, the volume is identified when both $d_h < N_h^r$ and $d_v < N_v^r$. When a violation of protection volume is identified, the interaction is computed using (3) or (6) depending on the type of uncertainty model considered. Since the violation of the protection volume can only occur when the points in question are in the same or in adjacent grid cells, the number of points to check is significantly smaller than in a pairwise comparison method.

In order not to underestimate interaction, one can simply choose a sufficiently small value of Δt. However, using small sampling-time step leads to large computation time and memory. Instead, we propose an inner-loop algorithm, detecting interaction between two sampling times, t and $t + \Delta t$, by *interpolating* aircraft positions with a sufficiently small step size, t_{interp}. Then, one checks each pair of these interpolated points. The algorithm stops when an interaction is identified or when every pair of the interpolated points have been checked. More details of this interaction detection algorithm are presented in [7].

5 Resolution Algorithm

To solve the strategic trajectory planning problem, we rely on a hybrid-metaheuristics approach adapted to handle an air traffic assignment problem at the continent scale. The proposed hybrid algorithm combines the simulated annealing (SA) and the local search (LS) algorithm such that the local search is considered as an inner-loop of the SA, which will be performed when a predefined condition is satisfied.

5.1 *Simulated Annealing*

Simulated annealing was separately introduced by S. Kirkpatrick et al. in 1982 [14] and by V. Černý in 1985 [17]. It is inspired by the annealing process in metallurgy where the state of a material can be modified by controlling the cooling temperature.

In the simulated annealing optimization algorithm, the objective function to be minimized is analogical to the energy of the physical problem, while the values of the decision variables of the problem are analogical to the coordinates of the material's particles. A control parameter, T, that decreases as the number of iterations grows plays the role of the temperature schedule, and a number of iterations, N_I, at each temperature step play the role of the time duration the material is kept at each temperature stage.

To simulate this evolution of the physical system toward a thermal equilibrium, the *Metropolis algorithm* [15] is used. For a given temperature, T, starting from a current configuration, the state space of the simulated system is subjected to a transformation (e.g., apply a local change to one decision variable). If this transformation improves the objective-function value, then it is accepted. Otherwise, it is accepted with a probability

$$P_{accept} := e^{\frac{\Delta E}{T}}, \tag{17}$$

where ΔE is the degradation of the objective-function value (negative for minimization). Repeating this process until the equilibrium is reached, the temperature is decreased according to a predefined cooling schedule. As the temperature decreases, the probability, P_{accept}, to accept a degrading solution becomes smaller and smaller. Therefore, the system will eventually converge to the nearest local optimum which will expectantly be close to a global optimum. We refer the reader interested by simulated annealing algorithm to the following books [11, 16].

For our problem, the simulated annealing proceeds as presented in [4]. In order to implement the simulated annealing algorithm to the strategic planning of 4D trajectories, we first define the following parameters.

1. **Neighborhood function.** To generate a neighborhood solution, first a flight i is randomly chosen. In order not to modify excessively the trajectories that are not involved in any interaction, we set a user-defined threshold value of interaction, denoted Φ_τ, such that the trajectory of a randomly chosen flight i will be modified only if

$$\Phi_i(u) \geq \Phi_\tau. \tag{18}$$

Otherwise, another trajectory will be randomly chosen until condition (18) is satisfied. This process ensures that changes will be first applied on trajectories involved in congestion area.

Then, for a chosen flight, i, we introduce a user-defined parameter, $P_w \leq 1$, to control the probability of modifying the value of the ith trajectory waypoint location vector, w_i, and a user-defined parameter $P_l \leq 1$ to control the probability of modifying the value of the flight level shift ΔFL_i. The probability to modify rather the departure time is therefore $1 - P_w - P_l$. These parameters, P_w and P_l, allow the user to set his/her preference on the way to deconflict trajectories. The neighborhood function we use in this paper is summarized in Algorithm 1.

Algorithm 1 Neighborhood function

```
Require: probabilities P_w, P_l, trajectory i.
Generate random number, r := random(0,1);
if r < P_w then
    Choose randomly one virtual waypoint w_i^m to be modified.
    Choose randomly new w_{ix'}^m from W_{ix'}^m;
    Choose randomly new w_{iy'}^m from W_{iy'}^m;
else
    if r < (P_w + P_l) then
        Choose randomly new flight level shift ł_i from ΔFL_i;
    else
        Choose randomly new departure time shift δ_i from Δ_i;
    end if
end if
```

2. **Initial temperature and initial acceptance probabilities.** To determine the initial temperature and initial acceptance probability, we rely on a practical recommendations given in [11]. They are computed by first generating 100 deteriorating transformations (neighborhood solutions) at random and then by evaluating the average variations, ΔE_{avg}, of the objective-function value. The initial temperature, T_0, is then deduced from the relation:

$$\tau_0 = e^{\frac{\Delta E_{avg}}{T_0}},$$

where τ_0 is the initial rate of accepting degrading solutions that will be empirically set.
3. **Cooling schedule.** The cooling schedule plays an essential role to guide the system toward a good optimum. If the temperature is decreased slowly, the system is more likely to converge to a better solution, but it will require more computation time. On the other hand, decreasing too rapidly the temperature tends to yield undesirable local optima. For simplicity, we will decrease the temperature, T, following the geometrical law, therefore

$$T_i = \beta . T_{i-1},$$

where $0 \leq \beta \leq 1$ where the constant β will be experimentally tuned.
4. **Equilibrium state.** In order to reach an equilibrium, a sufficient number of iterations, denoted N_I, or moves, have to be performed at each temperature step. For simplicity, the value of N_I will be defined as constant and will be experimentally set.
5. **Termination criterion.** Theoretically, it is suggested that the SA algorithm stops when the temperature reaches zero. However, this stopping criterion is not utilized in practice, since when the temperature is near zero, the probability of acceptance becomes negligible. In our case, the simulated annealing algorithm

will terminate when the final temperature, T_f, reaches the value $C.T_0$, where $0 \leq C \leq 1$ is a user-defined coefficient.

5.2 Hill-Climbing Local Search

Hill-climbing is a local search algorithm that only moves to a new solution only if it yields a decrease of the objective function. The process repeats until no further improvement can be found or until the maximum number of iterations $n_{T_{LOC}}$ is reached. In this work, we rely on two local search modules that correspond to the two following strategies:

- Intensification of the search on one particular trajectory (PT). Given a flight i, this state exploitation step focuses on improving the current solution by applying a local change from the neighborhood structure only to flight i.
- Intensification of the search on the interacting trajectories (IT). Given a flight i, this state exploitation step applies a local change, from the neighborhood structure to every flight that is currently interacting with flight i.

5.3 Hybrid Algorithm

In order to improve efficiency of the optimization algorithm in terms of computation time, we propose to combine the SA and the hill-climbing local search. The algorithms are combined in a self-contained manner, such that each algorithm is executed sequentially. The order of execution is controlled by predefined parameters that controls the probabilities to carry out each method. The probability to carry out simulated-annealing step, P_{SA}, is

$$P_{SA}(T) = P_{SA,min} + (P_{SA,max} - P_{SA,min}) \cdot \frac{T_0 - T}{T_0}, \tag{19}$$

where $P_{SA,max}$ and $P_{SA,min}$ are the maximum and minimum probabilities to perform the SA (predefined by the user). The probability of running a hill-climbing local search module, P_{Loc}, is given by

$$P_{Loc}(T) = P_{Loc,min} + (P_{Loc,max} - P_{Loc,min}) \cdot \frac{T_0 - T}{T_0}, \tag{20}$$

where $P_{Loc,max}$ and $P_{Loc,min}$ are the maximum and minimum probabilities to perform the local search (defined analogously). And, finally, the probability of carrying out both SA and the local search (successively), P_{SL}, is

$$P_{SL}(T) = 1 - (P_{SA}(T) + P_{Loc}(T)). \tag{21}$$

A key factor in tuning this hybrid algorithm is to reach a good trade-off between exploration (diversification) and exploitation (intensification) of the solution space, i.e., a compromise between fine convergence toward local minima and the computation time invested in exploring the whole search space.

6 Numerical Results

The proposed hybrid SA/LS algorithm is implemented in Java and run on an AMD Opteron 2 GHz processor with 128 Gb RAM. It is tested with two different uncertainty models, using real air traffic data at nationwide and continent scale.

6.1 Deterministic Uncertainty Model

First, the proposed algorithm is tested on national-size and continent-size air traffic, considering deterministic uncertainties model.

National-size air traffic data First, we test the proposed methodology on the full-day national-size en route air traffic over the French airspace involving 8,836 trajectories. Simulations are performed with different values for the parameters R_h, R_v, and t_ϵ, defining the size of the uncertainty sets. The parameter values chosen to specify the optimization problem are given in Table 1. The parameter values that specify the resolution algorithm are given in Table 2. The initial and final total interaction between trajectories, the computation time, and the number of iterations performed to solve the problems considering different levels of uncertainty is reported in Table 3 (the vertical uncertainty radius, R_v, is used only when aircraft are climbing and descending).

The size of the uncertainty set affects the resolution time and the final total interaction between trajectories. When increasing the time uncertainty, the initial interaction increases significantly (cases 1, 3, 4, and 5), and the algorithm requires more computation time to converge. The algorithm reaches an interaction-free

Table 1 Chosen (user-defined) parameter values specifying the robust optimization problem for the national-size air traffic

Parameter	Value
Discretization time step, Δt	20 seconds
Discretization time step for possible departure time shift, δ_s	20 seconds
Maximum departure time shift, $\delta_a^i = \delta_d^i := \delta$	120 minutes
Maximum allowed route length extension coefficient, d_i	0.20
Maximum allowed flight level shifts, $l_{i,max} := l_{max}$	2
Maximum number of virtual waypoints, M	3

Table 2 Empirically-set (user-defined) parameter values of the resolution methodology to solve the national-size air traffic

Parameter	Value
Number of iterations at each temperature step, N_I	200
Initial rate of accepting degrading solutions, τ_0	0.3
Geometrical temperature reduction coefficient, β	0.99
Final temperature, T_f	$(1/500).T_0$
Inner-loop interpolation sampling time step, t_{interp}	5 s
Probability to modify horizontal flight profile, P_w	1/3
Probability to modify flight level, P_l	1/3
Threshold value, Φ_τ	0.5 Φ_{avg}

Table 3 Initial and final total interaction between trajectories for the national-size air traffic, considering different dimensions for the deterministic uncertainty set

Case	Uncertainty set dimensions	Initial Φ_{tot}^D	Final Φ_{tot}^D	Solved interactions	CPU time (minutes)	No. of iterations
1	$R_h = 0$ NM. $R_v = 0$ ft. $t_\epsilon = 180$ s.	2,282,436	5,934	99.7%	1,093.8	1,083,215
2	$R_h = 1$ NM. $R_v = 100$ ft. $t_\epsilon = 60$ s.	765,448	0	100.0%	101.1	97,400
3	$R_h = 1$ NM. $R_v = 100$ ft. $t_\epsilon = 120$ s.	1,425,384	4,314	99.7%	1,809.0	1,791,000
4	$R_h = 1$ NM. $R_v = 100$ ft. $t_\epsilon = 240$ s.	2,821,706	37,290	98.7%	2,213.3	2,191,970
5	$R_h = 2$ NM. $R_v = 100$ ft. $t_\epsilon = 240$ s.	5,000,430	110,021	97.9%	2,289.8	2,266,956

solution for the case 2. It solves up to 99.7% of the initial interactions in the remaining cases (1, 3, 4, and 5), within computation times that are still compatible in a strategic planning context (the worst run, case 5, involving less than 38 h of CPU time).

Continent-size traffic data Then, the hybrid algorithm is tested on an air traffic data, involving en route air traffic over the European airspace. The data set is a full day of air traffic over the European airspace on 1st July 2011. It consists of 30,695 trajectories simulated with optimal vertical profiles and with direct routes. The user-defined parameter values specifying the optimization problem are the same as those given in Table 1. The maximum allowed flight level shift, $l_{i,max}$, is set to 0 due to the lack of data. The parameter values of the hybrid-metaheuristic algorithm are the same as those given in Table 2, with $N_I = 4{,}000$.

Table 4 Initial and final total interaction between trajectories for the continent-scale air traffic with different dimensions for the deterministic uncertainty set

Case	Uncertainty set dimensions	Initial Φ^D_{tot}	Final Φ^D_{tot}	Solved interactions	CPU time (minutes)	No. of iterations
6	$R_h = 3$ NM. $R_v = 200$ ft. $t_\epsilon = 60$ s.	5,142,632	634,474	87.7%	2,756.2	2,728,776
7	$R_h = 3$ NM. $R_v = 200$ ft. $t_\epsilon = 0$ s.	430, 234	0	100.0%	347.6	345,528

The initial and final total interaction between trajectories and the computation time to solve the problem considering different levels of uncertainty are reported in Table 4. Although the trajectories can be separated only by modifying the horizontal flight profile and the departure time of each flight, the resolution algorithm finds an interaction-free solution, taking into account uncertainty of aircraft positions, for problem instance in case 2. When time uncertainty is considered (case 1), there remains less than 15% of the initial interaction between trajectories.

6.2 *Probabilistic Uncertainty Model*

Then, the proposed robust strategic 4D trajectory planning methodology is tested based on the probabilistic-type uncertainty model. The parameters of the hybrid SA/LS are the same as those presented in Table 2. Again, the proposed algorithm is tested with the national-size air traffic over the French airspace. Assuming that aircraft is able to follow a given trajectory with high precision in the 3D space domain ($R_h = 0\,NM, R_v = 0$ ft.), the simulations are performed considering successively aircraft maximum time uncertainty, t_ϵ, of 1 up to 4 minutes, respectively. The initial and final interaction between trajectories and the required computation time are reported in Table 5. Remark that the initial total interactions between trajectories are significantly smaller than those of the worst-case-oriented approach. This is not surprising, since in the latter (deterministic) case, one counts one interaction, and in the former (probabilistic) case, there is even only a tiny positive probability of conflict.

The proposed strategic trajectory planning methodology is able to find interaction-free trajectory planning for all cases. When considering higher level of time uncertainty (4 minutes), the solution space becomes more constrained, and therefore the algorithm requires more computation time to converge.

Now we test the algorithm with the continent-size air traffic considering en route air traffic. The parameter values that specify the problem under consideration are, here again, the same as those given in Table 1. The parameters of the hybrid

Table 5 Numerical results for the national-size air traffic considering four different levels of aircraft maximum time uncertainty (1 to 4 min) based on probabilistic uncertainty model

t_ϵ (seconds)	Initial Φ^P_{tot}	Final Φ^P_{tot}	Solved interactions	CPU time (minutes)	No. of iterations
60	217,441.37	0.0	100.0%	116.07	114,970
90	274,953.55	0.0	100.0%	175.4	173,736
120	383,967.60	915.04	99.8%	586.3	1,031,730
240	718,374.42	1,547.13	99.8%	1,052.4	1,041984

Table 6 Numerical results for the continent-size instances, considering two different levels of time uncertainty based on probabilistic uncertainty model

t_ϵ (seconds)	Initial Φ^P_{tot}	Final Φ^P_{tot}	Solved interaction	CPU time (minutes)	No. of iterations
60	529,555.5	12,550.0	97.6%	1,341.7	1,328,152
120	1,079,738.4	40,706.2	96.2%	2,254.2	2,231,881

SA/LS are the same as those given in Table 2, with the number of iterations at each temperature step, N_I, empirically set to 2,000, more than for the above, smaller, national-size instance ($N_I = 200$).

The initial and final interactions between trajectories and computation time to solve the problem are reported in Table 6. Recall again that, as in the case without uncertainty, alternative flight levels for this continent-size instances are not available. Therefore, due to this lack of data, these problem instances can be separated only by modifying the horizontal flight profile and by modifying the departure time of aircraft. Nevertheless, there still remains less than 7% of the initial interactions taking into account the probabilistic-type time uncertainty.

7 Conclusions

In this paper, we have presented a methodology to solve 4D trajectory planning problem considering uncertainty of aircraft position and arrival time at strategic planning level. First, the uncertainties have been modeled with deterministic sets. The algorithm was tested on national-size and continent-size air traffic. To avoid being too conservative, probabilistic-type uncertainty sets were then considered.

The level of uncertainty to be considered is a trade-off between the desired robustness of the solution obtained and the associated trajectory modifications costs, to be decided by the user. Considering too important uncertainty in strategic planning will, indeed, result in a loss of capacity, since large portions of airspace have to be cleared for a given aircraft for a long period of time. Instead, the user can consider lower uncertainty levels and iteratively solve the remaining interactions during pre-tactical and tactical phases.

References

1. C. Allignol, N. Barnier, and A. Gondran. Optimized vertical separation in Europe. In *DASC 2012, the 31st IEEE/AIAA Digital Avionics Systems Conference*, pages 4B3-1–4B3-10, October 2012.
2. N. Barnier and C. Allignol. 4D - trajectory deconfliction through departure time adjustment. In *ATM 2009, the 8th USA/Europe Air Traffic Management Research and Development Seminar*, Napa (California), 2009.
3. N. Barnier and C. Allignol. Combining flight level allocation with ground holding to optimize 4D-deconfliction. In *ATM 2011, the 9th USA/Europe Air Traffic Management Research and Development Seminar*, Berlin (Germany), 2011.
4. S. Chaimatanan. *Planification Stratégique de Trajectoires D'avions*. PhD thesis, 2014.
5. S. Chaimatanan, D. Delahaye, and M. Mongeau. A methodology for strategic planning of aircraft trajectories using simulated annealing. In *ISIATM 2012, the 1st International Conference on Interdisciplinary Science for Innovative Air Traffic Management*, Florida, 2012. https://hal-enac.archives-ouvertes.fr/hal-00912772.
6. S. Chaimatanan, D. Delahaye, and M. Mongeau. Strategic deconfliction of aircraft trajectories. In *ISIATM 2013, the 2nd International Conference on Interdisciplinary Science for Innovative Air Traffic Management*, Toulouse (France), 2013. https://hal-enac.archives-ouvertes.fr/hal-00868450/document.
7. S. Chaimatanan, D. Delahaye, and M. Mongeau. A hybrid metaheuristic optimization algorithm for strategic planning of 4d aircraft trajectories at the continental scale. *Computational Intelligence Magazine, IEEE*, 9(4):46–61, Nov 2014.
8. S. Constans, B. Fontaine, and R. Fondacci. Minimizing potential conflict quantity with speed control. In *The 4th Eurocontrol Innovative Research Workshop And Exhibition*, pages 265–274, Bretigny-sur-Orge (France), 2005.
9. N. Dougui, D. Delahaye, and M. Mongeau. A new method for generating optimal conflict free 4D trajectory. In *The 4th International conference on research in air transportation*, pages 185–191, Budapest (Hungary), 2010.
10. N. Dougui, D. Delahaye, S. Puechmorel, and M. Mongeau. A light-propagation model for aircraft trajectory planning. *Journal of Global Optimization*, 56(3):873–895, 2013.
11. J. Dreo, A. Petrowski, P. Siarry, and E. Taillard. *Metaheuristics for hard optimization*. Springer, 2006.
12. N. Durand and J. B. Gotteland. Genetic algorithms applied to air traffic management. In *Metaheuristics for Hard Optimization*, pages 277–306. Springer, 2006.
13. M. R. Jardin. *Towards Real Time En Route Air Traffic Control Optimization*. PhD thesis, Standford University, 2003.
14. S. Kirkpatrick, C. D. Gelatt, and M. P. Vecchi. Optimization by simulated annealing. *Science*, 220:671–680, 1983.
15. N. Metropolis, A. W. Rosenbluth, M. N. Rosenbluth, A. H. Teller, and E. Teller. Equation of state calculations by fast computing machines. *Journal of Chemical Physics*, 21:1087–1092, 1953.
16. E.-G. Talbi. *Metaheuristics: From Design to Implementation*. Wiley, 2009.
17. V. Černý. Thermodynamical approach to the traveling salesman problem: An efficient simulation algorithm. *Journal of Optimization Theory and Applications*, 45(1):41–51, 1985.

Aircraft Trajectory Planning by Artificial Evolution and Convex Hull Generations

S. Pierre, D. Delahaye, and S. Cafieri

Abstract Air Traffic Management (ATM) ensures the safety of flights by optimizing flows and maintaining separation between aircraft. Many ATM applications involve some aircraft trajectory optimization in order to improve the performance of the overall system. Trajectories are objects belonging to spaces with infinite dimensions. Widely used approaches are based on discretization, sampling trajectories at some regular points, and then using appropriate representations to reduce the dimension of the search space. We propose an approach in which trajectories in a two-dimensional space are designed with the help of convex hull generation. By using static as well as moving obstacles for which the position and the size are controlled by artificial evolution, we propose a new algorithm for efficient trajectory planning in Terminal Maneuvering Areas.

Keywords Trajectory planning • TMA • Optimization • Obstacle avoidance

1 Introduction

Trajectory planning is crucial in Air Traffic Management (ATM) to regulate air traffic flows while ensuring flight safety. Trajectories are designed so as to avoid aircraft potential conflicts as well as to optimize some criteria, such as cost index or environmental criteria (noise abatement, pollutant emission, etc...). Depending on the considered time horizon, the following kinds of trajectory planning can be carried out:

- at a strategical level, only macroscopic indicators like congestion, mean traffic complexity, delays can be taken into account, as well as the presence of obstacles;
- at a pre-tactical level, the accuracy of previous indicators, specially congestion and complexity, increases, while at the same time, early conflict detection can be performed;

S. Pierre • D. Delahaye (✉) • S. Cafieri
Applied Mathematics Laboratory, French Civil Aviation University, Toulouse, France
e-mail: spierre@etud.insa-toulouse.fr; delahaye@recherche.enac.fr; sonia.cafieri@enac.fr

Electronic Navigation Research Institute, *Air Traffic Management and Systems II*,
Lecture Notes in Electrical Engineering 420, DOI 10.1007/978-4-431-56423-2_3

- finally, at the tactical level, conflict resolution is the major concern and optimality of the trajectories is only marginally interesting.

In this work, we focus on path planning in Terminal Maneuvering Areas (TMAs), which are the areas surrounding one or more neighboring airports, where arriving and departure routes (also called STAR and SID, respectively) have to be handled. Such a planning is done at a strategical level and aims at designing trajectories avoiding obstacles. In this paper we propose also an extension of the algorithm in order to design "dynamic" SID and STAR in case of moving obstacles such as weather. This more advanced concept of dynamic paths ensures to maintain flows to the runways and avoid vectoring in TMAs which is a critical issue in those areas.

An example of such routes is given in Fig. 1 which represents the New York area.

The paths flown by aircraft are considered as curves in $\mathbb{R}^3$. Such time-independent trajectories are called *shapes*.

Aircraft trajectories are usually designed into three steps. The first step consists in the design of the two-dimensional shape between two points, respectively, the origin and the destination point. Then, an optimal altitude profile is computed in order to create a full three-dimensional shape. Finally, the speed profile is computed in order to optimize some cost criteria (fuel, cost index). When we consider departure

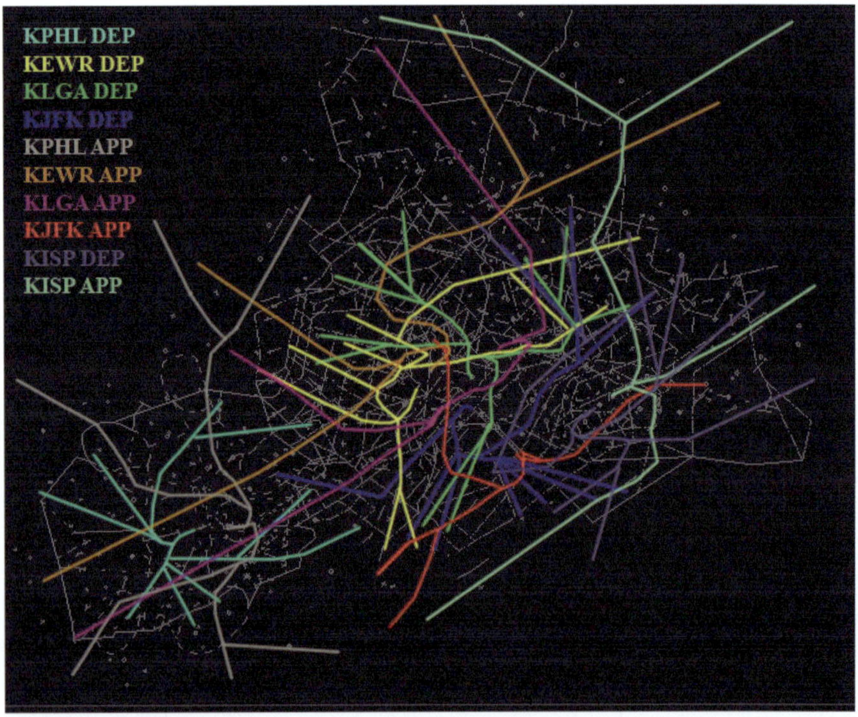

Fig. 1 Departure and arrival routes for the major airports in the vicinity of New York

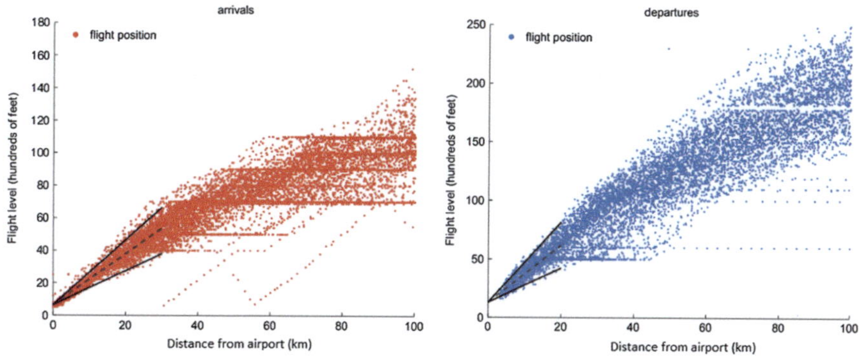

Fig. 2 Departure and arrival trajectory samples for Roissy Airport

or arrival route design, such design has to be able to address any kind of altitude profiles. As a matter of fact, the climbing (descending) rate of an aircraft depends strongly on the type of aircraft (B747, A320, etc...), its weight, the headwind (or tailwind), and the outside temperature. For instance, recorded departure and arrival trajectory samples for Roissy Airport are given in Fig. 2. As it can be seen on the figure, climbing (descending) rates undergo large deviations.

In this work, we propose a new approach to optimize the two-dimensional shape of an aircraft trajectory.

So, the obtained 2D shapes are supposed to be evaluated in the three-dimensional space by using some given altitude profiles.

The paper is organized as follows: Sect. 2 outlines the main existing approaches for trajectory planning. Section 3 presents the proposed model and algorithm. First, the trajectory shape design is introduced, and then a combinatorial optimization problem and an evolutionary algorithm for its solution, used to carry out such a design, are presented. Section 4 presents an extension of the proposed approach to take efficiently into account the case of dynamic obstacles. Some results validating such an approach are discussed in Sect. 5. Finally, Sect. 6 draws some conclusion.

2 State of the Art

Trajectories are mathematical objects belonging to spaces with infinite dimensions. Widely used approaches for their design are based on discretization. The general ideas are to use a discrete number of parameters describing the trajectory, optimize such parameters with respect to some selected criterion to design a given class of trajectory shapes, and finally build a trajectory γ. This process is summarized in Fig. 3. Starting from a discrete set of points, a way to build a trajectory γ is based on using interpolation. Piecewise linear interpolation is the simplest piecewise interpolation method. An example of such a linear piecewise interpolation is shown

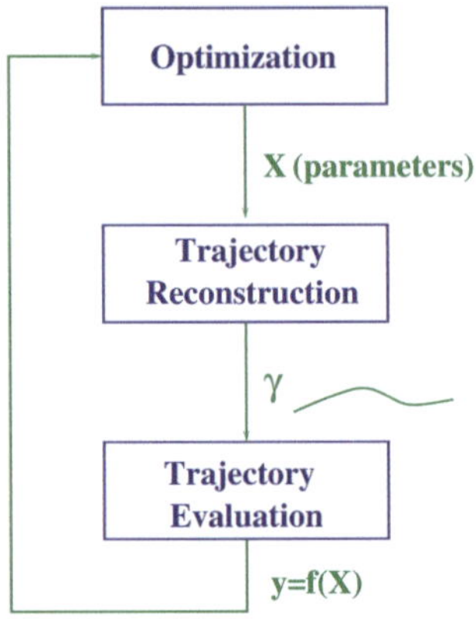

Fig. 3 The optimization process controls the X vector in order to build a trajectory γ for evaluation

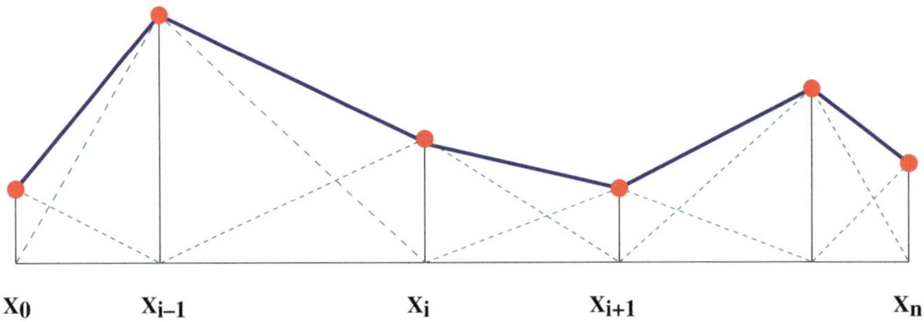

Fig. 4 Piecewise linear interpolation. By summing the triangle shapes (*dash lines*) controlled by the positions of the *red dot* and the associated extension, we can generate a full linear piecewise interpolation (*solid line*)

in Fig. 4. In this case, the derivative of the resulting curve is not continuous. In order to fix this drawback, one can use piecewise quadratic, piecewise cubic interpolation [7] or cubic spline interpolation [2], which ensures smooth C^2 shapes. Alternatively, when interpolating the given points is not a hard constraint, one can use some control points which control the shape of a given trajectory without forcing this trajectory to go through those points. Among such kind of approximation approaches, we may recall the ones based on Bézier curves [4] or on B-splines [3]. If many points have to be considered, using a Bézier curve, one has to manipulate polynoms with high degrees. B-splines allow to circumvent this weak point. A B-spline is a spline function that has minimal support with respect to a given degree, smoothness, and domain partition. An example of B-spline of order 1 is given in Fig. 5. To increase the smoothness of the resulting shape, one may use B-splines of order 3. When many aircraft trajectory samples are available (for instance, from radar), one can build a dedicated basis and minimize the number of coefficients for

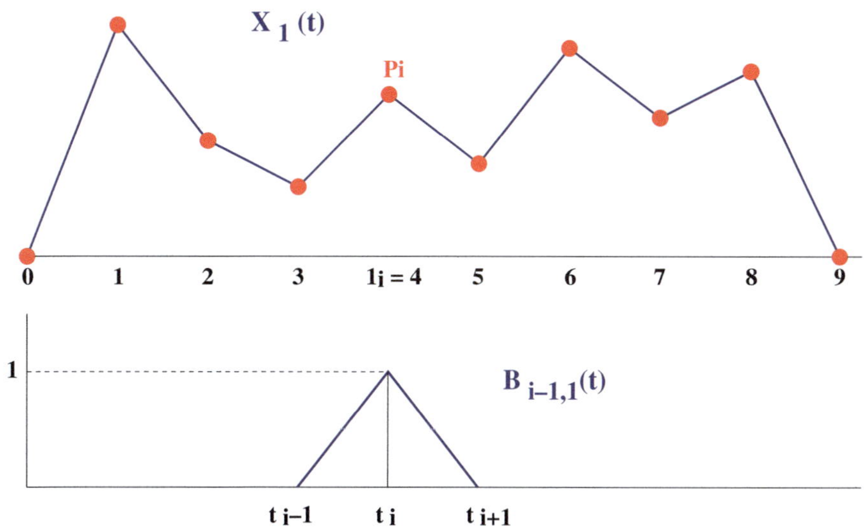

Fig. 5 An example of B-spline of order 1

trajectory reconstruction:

$$\gamma_i(t) = \sum_{k=1}^{k=K} a_{ik}\psi_k(t) \tag{1}$$

For instance, principal component analysis (PCA [5]) can be used to convert a set of observations of possibly correlated variables into a set of values of linearly uncorrelated variables, called principal components.

Another easy way to build trajectory is to use some reference trajectories (regular trajectories used by aircraft) and to compute a weighted sum of such reference trajectories to build a new one. Considering two (or more) reference trajectories γ_1, γ_2 joining the same origin-destination pair (see Fig. 6) (past flown trajectories may be considered), one can create a new trajectory γ_α by using an homotopy:

$$\gamma_\alpha = (1-\alpha)\gamma_1 + \alpha\gamma_2, \quad \alpha \in [0, 1]. \tag{2}$$

Remark that the above approaches do not take into account obstacles in the design process. Obstacles can nevertheless be included as a penalty in the objective function. Trajectory design in a constrained space (i.e., with obstacles) has been looked at through various techniques. They include the A* algorithm to provide obstacle-free trajectories, followed by the use of some smoothing algorithm to improve their regularity [8]. Another approach consists in using a branch and bound algorithm where the branching strategy is associated to the way obstacles are avoided, bypassing them in one direction or the other [1].

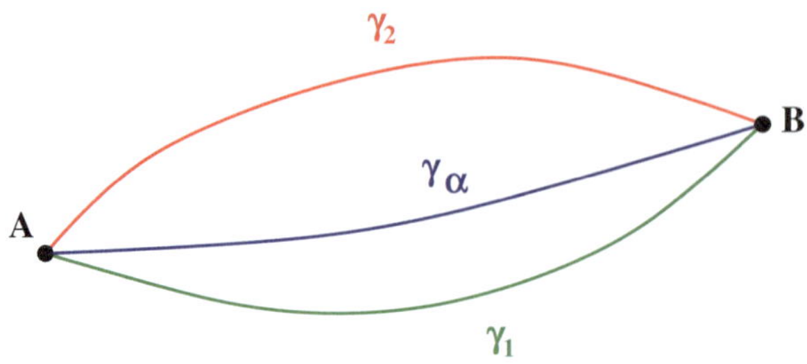

Fig. 6 An example of B-spline of order 1

In this paper we propose an approach based on obstacle convex hull generation in order to create paths around some given sets of obstacles. Two sets of obstacles are then considered: one of true obstacles and one of virtual obstacles. The proposed model is presented in the next section.

3 Optimization Model and Algorithm

3.1 Trajectory Shape Design

Let us consider, in a two-dimensional space, two points representing the origin and the destination of the trajectory to be designed (the origin being the point with the smallest x-axis value) and a set Ω of obstacles. Let us model obstacles in Ω by sets of points defining their 2D contour shapes. These points have x-axis values between those of the origin and destination. We suppose these points defining convex sets; if this is not the case, we consider the associated convex envelope as the obstacle instead of the original one. We also suppose that the points defining obstacles are ordered clockwise or counterclockwise.

Building on the technique from R.A. Jarvis, known as *Jarvis March* [6], we propose a trajectory planning approach based on building convex hulls around obstacles.

Specifically, we use the following adaptation of the Jarvis March or gift-wrapping algorithm. Let G be a set of two-dimensional points which represent obstacles. We first consider the point M_0 such that

$$\forall P \in G,\ (P)_x \geq (M_0)_x \tag{3}$$

where $(P)_x$ denotes the x-value of point P. We then look for the next point M_1 such that

$$\forall P \in G,\ P \text{ is on the left side of } \overrightarrow{M_0 M_1}$$

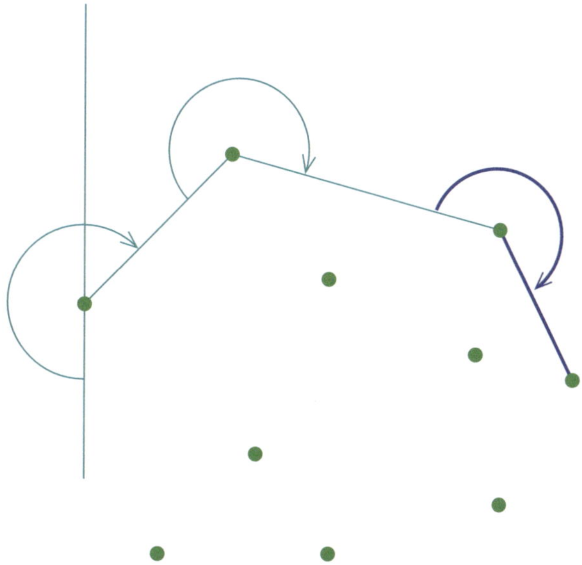

Fig. 7 The algorithm first classifies point from *left* to *right*. Then, it considers the point at extreme *left* and opens an angle from this point with the y direction. This angle is increased until the associated segment crosses a new point. The process is iterated from this new point until the algorithm reaches the first point again. All the points selected by this process make the searched convex hull

Iterating this procedure, we get the point $M_{n+1} \in G$ such that

$$\forall P \in G, \ P \text{ is on the left side of } \overrightarrow{M_n M_{n+1}}$$

We stop once M_{n+1} is equal to M_0. The set $C = \{M_0, \ldots, M_n\}$ is the so-called convex hull of G.

The original Jarvis algorithm is illustrated in Fig. 7.

The convex hull created through the above technique, in the case of two obstacles, describes two paths around the obstacles like illustrated in Fig. 8. Remark that the described algorithm is not able to generate shapes like the one shown in Fig. 9. To build a trajectory with a shape like the one shown in Fig. 9, which is more close to what is done in practice in the operational context and whose length is less than the one of a trajectory obtained with the above algorithm, we propose a variant of the algorithm which generates piecewise convex hull segments. To this aim, we first partition the set of obstacles into two subsets: the set O_a of obstacles to be avoided by a "convex" trajectory and the set O_b of obstacles to be avoided with a "concave" trajectory. Let G_a (respectively, G_b) be the set of points describing O_a (respectively, O_b). Similarly to what is done when the Jarvis March algorithm is used, we first consider the point M_0 such that

$$\forall P \in G_a, \ (P)_x \geq (M_0)_x \tag{4}$$

Then, proceeding iteratively, we define the point M_{n+1}:

$$\forall P \in G_a, \ P \text{ is on the left side of } \overrightarrow{M_n M_{n+1}} \tag{5}$$

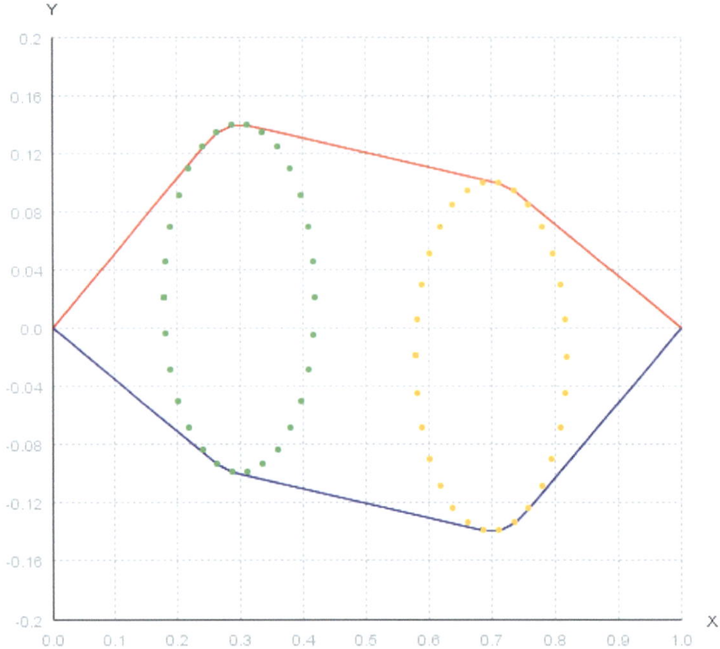

Fig. 8 Trajectories around two obstacles, built by using the Jarvis March algorithm. Two segments join the extreme points

The difference with the above Jarvis March algorithm lies in the stopping criterion. This time we stop the algorithm when the following condition is satisfied:

$$\forall P \in G_b,\ P \text{ is on the right side of } \overrightarrow{M_n M_{n+1}} \text{ with} \qquad (6)$$
$$M_n, M_{n+1} \in G_a$$

The point M_{n+1} is not included in the convex hull C_a. We get a kind of semi-convex hull. An example of trajectory around two obstacles obtained by the proposed algorithm is given in Fig. 10. Note that the above algorithm only builds the blue part of the trajectory in Fig. 10. To produce the red part, the algorithm has to be applied again with O_b and another obstacle O_c. This time, since we want to bypass the obstacle through a concave trajectory, we follow exactly the same procedure except that all the *right* − *left side* comparisons are reverted, i.e., the stopping criterion is changed to

$$\forall P \in G_c,\ P \text{ is on the left side of } \overrightarrow{M_n M_{n+1}} \text{ with} \qquad (7)$$
$$M_n, M_{n+1} \in G_b$$

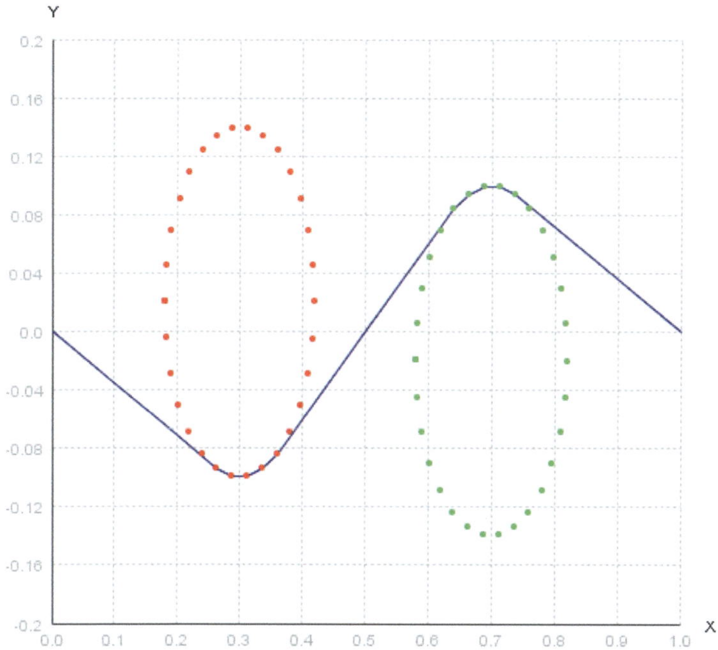

Fig. 9 Trajectory around two obstacles. It is built with two segments, one convex and another one concave

To connect the two pieces of trajectories created as described above, we just add the last point of C_a to G_b. Finally, a full trajectory is built by considering the starting point S as the first obstacle and the ending point E as the last obstacle, corresponding to the origin and the destination. Such a trajectory is like the one in Fig. 9. We can now describe our optimization model.

3.2 Decision Variables

Let Ω be the set of all n obstacles to be avoided. Let O_p and $O_q \in \Omega$ be two obstacles defined by the set of points G_p and G_q:

$$p < q \Rightarrow \left(M_p\right)_x \leq \left(M_q\right)_x \tag{8}$$

with

$$\left(M_p\right)_x \leq (P)_x, \forall P \in G_p \tag{9}$$

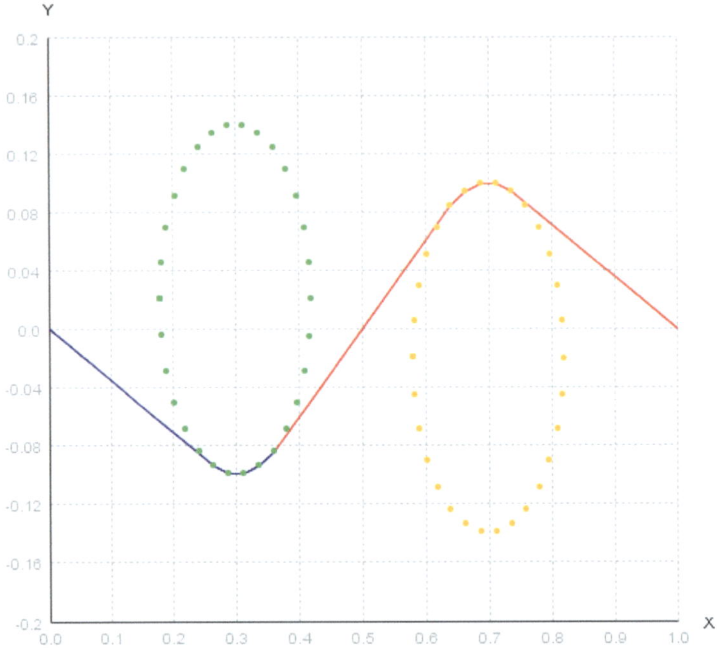

Fig. 10 Trajectory around two obstacles. The new version of the Jarvis March algorithm is used

and

$$\left(M_q\right)_x \leq (P)_x, \forall P \in G_q \tag{10}$$

We consider obstacles ordered from the "left to the right." The decision variables of our optimization model are binary variables defined as follows:

$$X_i = \begin{cases} 1 \text{ if obstacle } i \text{ is to be bypassed} \\ 0 \qquad\qquad \text{otherwise} \end{cases} \tag{11}$$

and

$$Y_i = \begin{cases} 1 \text{ if obstacle } i \text{ is to be bypassed by the bottom} \\ \qquad \text{(convex piece of trajectory)} \\ 0 \quad \text{if obstacle } i \text{ is to be bypassed by the top} \\ \qquad \text{(concave piece of trajectory)} \end{cases} \tag{12}$$

The space state of the optimization problem is then defined by $2 \times n$ variables, with n the total number of obstacles. Figure 11 gives an example of decision variable values where two obstacles have to be avoided.

	X	Y
Obstacle 1	1	1
Obstacle 2	1	0

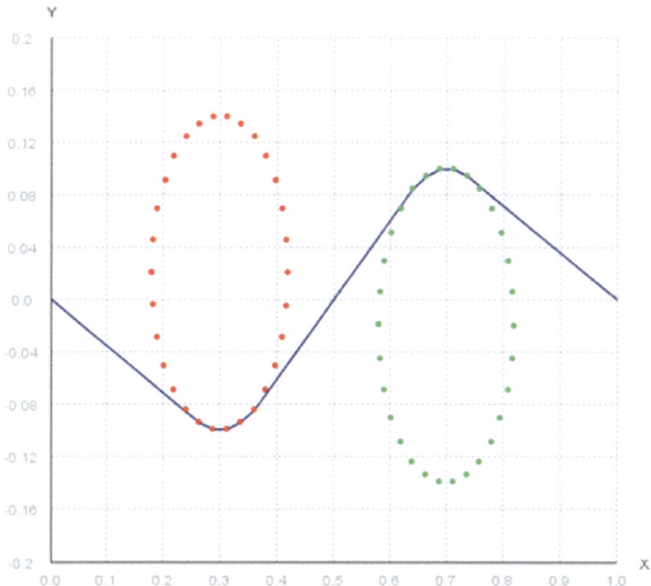

Fig. 11 In this example both obstacles are active; the second one has to be avoided by the top

3.3 Objective Function

The objective function, to be minimized, is composed of two parts, each accounting for a criterion to be taken into account. The first one is linked to the length of the (discrete) trajectory T:

$$L(T) = \sum_{i=1}^{n} l(P_i, P_{i-1}) \tag{13}$$

with $T = \{P_i, i \in \{1, \ldots, n\}\}$ and $l(M, N)$ the distance between the two points M and N. The second one is associated to pieces of trajectories going through obstacles: even if a piece of trajectory crosses an obstacle in the two-dimensional space, it may happen that it does not intersect the obstacle on the vertical plan. This situation can be detected when the three-dimensional space is considered, in a second step, after that the two-dimensional shape of the trajectory is obtained. We introduce the following function:

$$D(T) = \sum_{i=1}^{m} \sum_{j=1}^{n} R_i \times \Delta(P_j, O_i) \tag{14}$$

where R_i is the cost for going inside the obstacle O_i, and $\Delta(P, O) = 1$ if the point P is inside the obstacle O and $\Delta(P, O) = 0$ otherwise. To detect if a point is or not inside an obstacle, we use the assumptions that obstacles are convex and the fact that the points defining obstacles are ordered in the clockwise direction. Then:

$$P \text{ is in } O_p \Leftrightarrow (P, M_1)(P, M_2) \geq 0, \forall M_1, M_2 \in G_p$$

where $P \in \mathbb{R}^2$, O_p is defined by the set of points G_p and (P, M) is the cross product between P and M.

Finally, the objective function of our optimization model is defined as follows:

$$F(T) = L(T) + D(T) \tag{15}$$

and has to be minimized.

3.4 Genetic Algorithm

The problem to be solved is a combinatorial optimization problem, whose complexity is directly related to the number of obstacles. We then address the problem by using an artificial evolution algorithm, where a stochastic tournament selection process is used associated with an elitism principle. More specifically, we use a genetic algorithm, where three kinds of mutation operators, $\psi_i\, i = 1 \ldots 3$, are applied with a given probability. The first mutation operator (ψ_1) randomly generates two new vectors X and Y. It is very disruptive and is mainly applied at the beginning of the evolution process. The second operator (ψ_2) randomly changes a percentage of bit in X and Y (this percentage is diminishing with the generation number). The third one (ψ_3), in the case of a trajectory crossing some obstacles, randomly chose an obstacle O_p between the ones which are crossed by the trajectory and change the values of X_p and Y_p. Crossover operators have also been developed, but from some numerical tests, it appears that they do not really improve the performance of the algorithm, so only mutation operators are kept.

4 Model Extension

4.1 Adding a Time Dimension

4.1.1 New Obstacle's Definition

Aircraft trajectory design is also subject to dynamic obstacles like storms moving in the airspace which have to be avoided. We then extend the previous approach so as to take into account dynamic obstacle avoidance. We consider obstacles in three dimensions, including now the time dimension in their definition. So, we define obstacles as discrete structures composed by a starting time and an ending time, a set of points in two-dimensional space, and a speed vector. Let O_p be a three-dimensional obstacle. Let t_o be the starting time, t_f the ending time, G_o the set of points at t_o, and v the speed vector. Then, O_p is defined by $O_P = \{t_o \in \mathbb{R}^+, t_f \in \mathbb{R}^+, G_o, v \in \mathbb{R}^2\}$. Let $G(t)$ define the set of points representing the obstacle at time t:

$$G(t) = \begin{cases} \emptyset \ , t \notin \left[t_0, t_f\right] \\ \{P + v \times t, P \in G_o\} \ , t \in \left[t_0, t_f\right] \end{cases} \tag{16}$$

The current framework enables static shapes for obstacles, but this model can be adapted in order to manage obstacles with shapes changing with time. In this case we have to consider $O_p(t)$ instead of O_p.

4.1.2 Obstacles Intersection

In order to dynamically detect when a generated trajectory crosses a given moving obstacle O_p, we use the pretty simple following assertion.

Let $P = (x, y, t) \in \mathbb{R}^2 \times \mathbb{R}^+$ be a point,

$$P \in O_p \Leftrightarrow \begin{cases} t_o \leq t \leq t_f \\ (P, M_1(t))(P, M_2(t)) \geq 0, \ \forall M_1(t), M_2(t) \in G_p(t) \end{cases} \tag{17}$$

This corresponds to the spatial-time extension of the obstacles.

4.2 Solution Approach

4.2.1 First Approach

Once obstacles are defined including their time dimension, an issue arises in computing a convex hull in order to extract two unique paths. Indeed, to do so, we need to consider obstacles in a two-dimensional space and design trajectories around

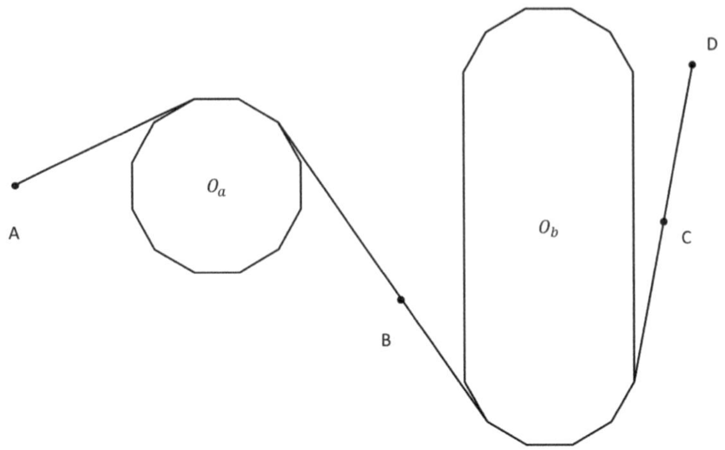

Fig. 12 Trajectory created using the whole projection of O_b in the two-dimensional space

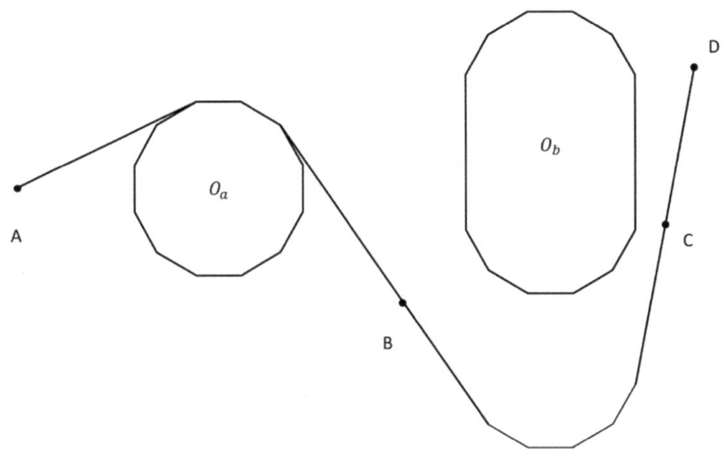

Fig. 13 Projection of O_b restricted to the traveling time between points B and C

them. However, it is not an easy task to determine an obstacle position when the obstacle is moving. We can, for example, use a pessimistic approach, which consists in avoiding an obstacle on all the space that he would cover while we go from a point A to a point D designing the trajectory (see Figs. 12 and 13). In the example in Fig. 13, O_a is not moving and O_b has the same shape as O_a but is moving from the bottom to the top. However, when we are around the obstacle O_b (in other words, when the aircraft flying on the trajectory is on the section between the points B and C), the obstacle has moved and the previous planned trajectory (using the projection of the obstacle) is no more optimal. To solve this issue, we dissociate the trajectory definition in the two-dimensional space (still building the trajectory around two-dimensional obstacles) from the evaluation through the objective function, which

Table 1 Dimension of the state space

Time step	Ω	Ω_m
0.5	5.904×10^4	3.874×10^8
0.2	5.904×10^4	2.058×10^{14}
0.1	5.904×10^4	7.178×10^{23}

takes into account the other dimensions. The problem becomes how to transcript our three-dimensional elements into two-dimensional elements in an efficient way. A first way to address this problem is transforming a moving object into a large number of static objects existing only for a short period of time all over the trajectory of the moving object. We let the genetic algorithm decide which one of these static objects is the best one to consider. However, doing this increases dramatically the computation time to solve our problem. To figure out this problem, let Ω be a set of obstacles composed of ten static obstacles. Let O_m be a moving obstacle and $\Omega_m = \Omega \cup \{O_m\}$. Obstacle O_m appears at $t_o = 1.0$ s and disappears at $t_f = 5.0$ s. In Table 1, we report the dimension of the state space with the associated time step. To compute these values, we assume that each point of the state space is composed of n variables, where n is the number of obstacles. Each of these variables can take three values, depending on the choice of ignoring the obstacle and go through it, bypassing it in a concave way or bypassing it in a convex way. We obtain 3^n possibilities for each point of the state space. The number n when there are moving obstacles depends on the time step. Since in our example O_m exists during a 4 s period, if the time step is 0.5, we will need 8 obstacles to represent O_m. Let N be the number of possibilities for a chromosome; we have

$$N = 3^{n+\frac{t_f - t_o}{\Delta t}} \tag{18}$$

where $n = card(\Omega)$ and Δt is the time step.

4.2.2 Cluster Approach

As an alternative to the above approach, we propose an approach based on clusters of obstacles. We define a cluster as a set of n obstacles, where one randomly selects a number $m \leq n$ of obstacles to be bypassed at a given time, ignoring the others. Note that a cluster contains possible positions for the same obstacle at different times. Furthermore, an aircraft following the designed trajectory will be close to the obstacle between two times t_a and t_b, so the position of the obstacle at these two times can be used to build the trajectory. The values of times t_a and t_b are computed using the genetic algorithm presented above. The algorithm will determine the obstacles in clusters and the associated times t_a, t_b (for each of them). Based on the performance of the associated choices, the evolution process will adjust the members of the clusters and the associated selected times in order to improve the overall fitness.

Table 2 Comparison of the cluster and the naive approach in terms of size of the state space

Time step	Cluster method	Naive approach
0.5	1.134×10^7	3.874×10^8
0.2	7.086×10^7	2.058×10^{14}
0.1	2.834×10^8	7.178×10^{23}

Remark that for static obstacles a cluster is defined as a set composed of only one obstacle with $m = 1$, which corresponds to the simple two-dimensional case.

The number of possible combinations N corresponding to a system of n clusters is given by

$$N = 3^n \prod_{i=1}^{n} r_i^{m_i} \tag{19}$$

where

n is the number of clusters.
r_i is the number of obstacles in the ith cluster.
m_i is the number of obstacles that can be used at the same time in the ith cluster.

In Table 2, the dimensions of the state space in the above naive approach and in the cluster approach are compared, showing the drastic size reduction induced by the latter.

4.2.3 Genetic Algorithm: New Chromosome Definition

Using the cluster-based approach, we need to change the meaning of the variables in a chromosome, to be used in the genetic algorithm. Indeed, in this case we do not control exactly which obstacle we are going to avoid, but we control which cluster we are going to avoid. Let Ω be a set of n obstacles that form k clusters. Let C be a cluster and G_c the set of obstacles included into C. Let m be the number of obstacles in C to be possibly avoided. When C is activated during the computation (at the initialization phase or after a mutation), the genetic algorithm chooses randomly m obstacles within G_c and uses them to construct the trajectory. The structure of the new chromosome is given in Fig. 14.

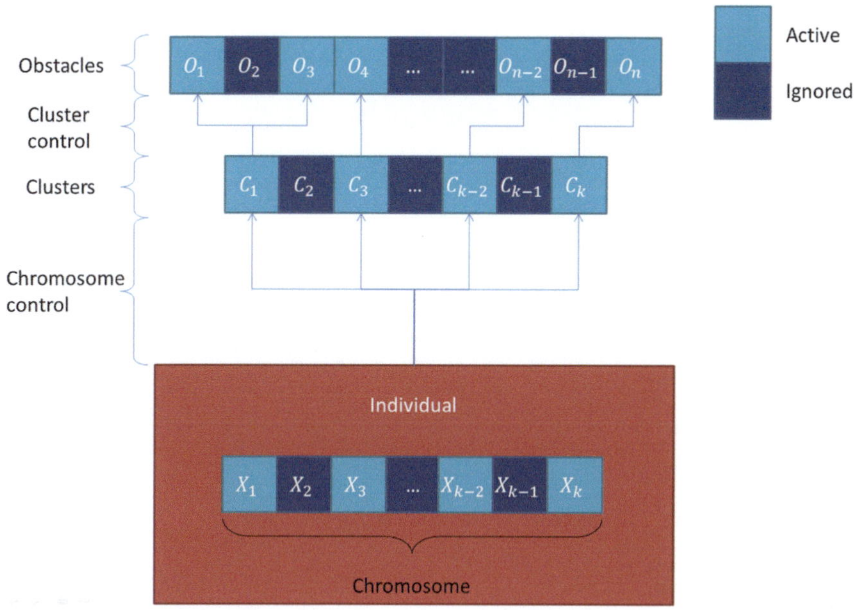

Fig. 14 Management of obstacles using clusters. The dimension of the chromosome is strongly reduced with respect to the first proposed approach, while the performance of the algorithm is nearly the same

5 Some Results with Moving Obstacles

Let consider an instance of the problem of designing a trajectory avoiding static and moving obstacles, like the one depicted in Fig. 15. In this example there are 18 obstacles, which are all static but one, represented in blue color, which models a moving storm. The storm appears on the left side and moves to the right. During its movement, it describes a shape that it covers at different times. Suppose that we want to design a trajectory from the origin point $(0.0, 2.5)$ to the destination point $(4.2, 2.5)$. To do so, we apply the above genetic algorithm with the parameters in Table 3. The computing time to obtain the solution is about 20 s on a Pentium 3 GHz. The trajectory designed by the algorithm is shown in Fig. 16. This trajectory is based on intermediate states of the storm to avoid it. Displaying an animated aircraft moving along the designed trajectory, we indeed see that it avoids the storm almost perfectly.

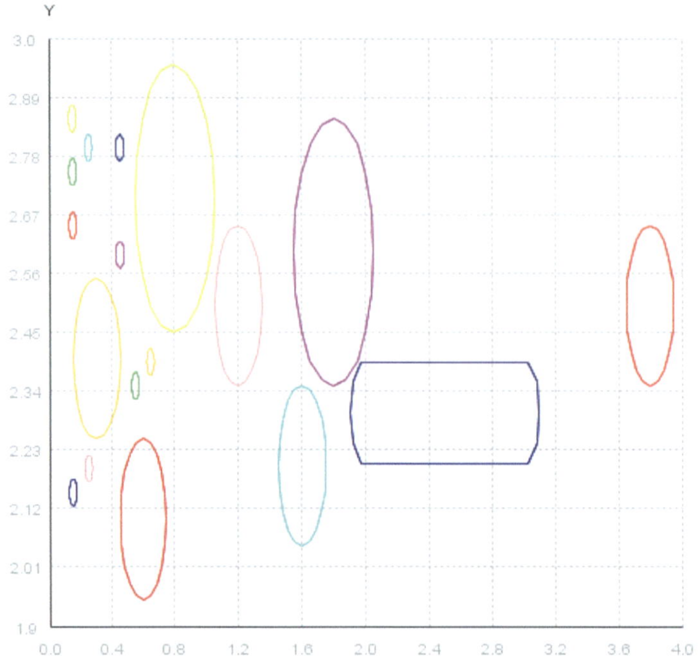

Fig. 15 Obstacles in Ω. The *blue* obstacle has been drawn for all the duration on its appearance

Table 3 Genetic algorithm: parameters

Population size	500
Number of generations	500
Probability of mutation ψ_1	0.20
Probability of mutation ψ_2	0.15
Probability of mutation ψ_3	0.65

6 Conclusion

We presented a trajectory planner which is able to perform obstacle avoidance in an efficient way. We first adapted the Jarvis March algorithm in order to generate convex or concave obstacle avoidance and have used it to develop a trajectory generation in a two-dimensional space that avoids obstacles while optimizing the trajectory length. The problem is modeled as a combinatorial optimization problem, whose size may be large, especially in the case of a large number of obstacles and in that of moving obstacles. Thus, we propose to address the problem with an artificial evolution approach. Specifically, we propose a genetic algorithm, where mutation operators have been developed with different granularities and applied with different probabilities. An extension based on clusters of obstacles has also been developed in order to improve the performance of the algorithm in the case of moving obstacles.

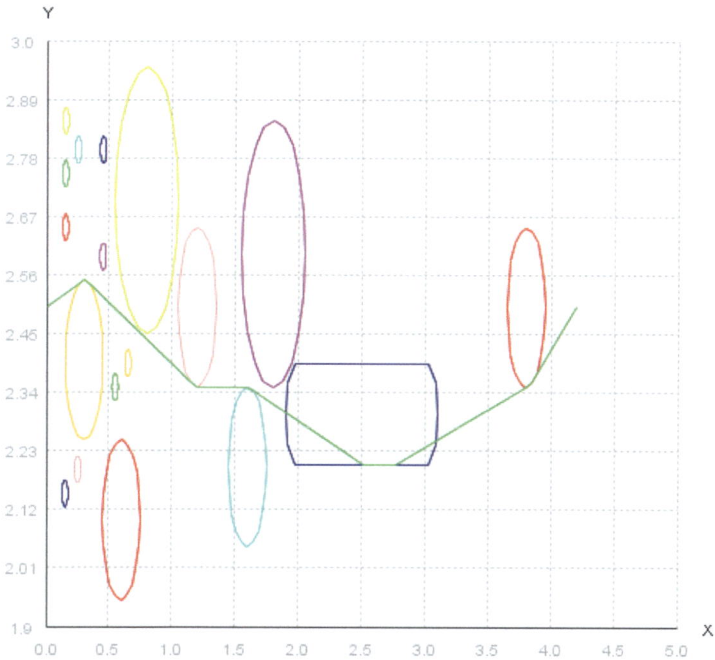

Fig. 16 Designed trajectory. Remind that the *blue* obstacle has been drawn for its whole time extension. The aircraft does not intersect it at any time

Future work will address the extension of the proposed approach for the design of trajectory shapes in a three-dimensional space, to address more closely the design of departure and arrival 3D routes in Terminal Maneuvering Areas.

Acknowledgements This work has been partially supported by French National Research Agency (ANR) through grant ANR 12-JS02-009-01 ATOMIC.

References

1. Eele A. and Arthur R., *Path-planning with avoidance using nonlinear branch-and-bound optimization*, Journal of Guidance Control and Dynamics **32** (2009), no. 2, 384–394.
2. Birkhoff and C de Boor, *Piecewise polynomial interpolation and approximation.*, Proceeding of the General Motors Symposium of 1964, General Motors, 1964.
3. C de Boor, *A practical guide to splines*, Springer-Verlag, 1978.
4. Farin G. and Hansfordman D., *The essentials of cagd*, A. K. Peters, Ltd., 2000.
5. Abdi H. and Williams L.J., *Principal component analysis*, Interdisciplinary Reviews: Computational Statistics **2** (2010), 433–459.
6. Jarvis R.A, *On the identification of the convex hull of a finite set of points in the plane.*, Information Processing Letters **2** (1973), no. 1, 18–21.
7. M. T Hearth, *Scientific computing, an introductory survey*, Computer graphics. McGraw-Hill, 2002.
8. Zhao Y. and Tsiotras P., *A quadratic programming approach to path smoothing.*, Proceeding of the American Control Conference (ACC), ACC, 2011.

Homotopy Route Generation Model for Robust Trajectory Planning

Andrija Vidosavljevic, Daniel Delahaye, and Vojin Tosic

Abstract Although advance future avionics will enable full compliance with the given trajectory, there are many uncertainty sources that can deflect aircraft from their intended positions. In this article, we investigate potential of robust trajectory planning, considered as an additional demand management action, as a means to alleviate the en route congestion in airspace. Robust trajectory planning (RTP) involves generation of congestion-free trajectories with minimum operating cost taking into account uncertainty of trajectory prediction and unforeseen event. The model decision variables include ground delay, change of horizontal route, and vertical profile (flight level) to resolve congestion problem. The article introduces a novel approach for route generation (3D trajectory) based on homotopic feature of continuous functions. It is shown that this approach is capable of generating a large number of route shapes with a reasonable number of decision variables. RTP problem is modeled as a mixed-variable optimization problem, and it is solved using stochastic methods. The model is tested on a real-life example from the French airspace. The results indicate that, under certain conditions, at the expense of a small increase of total planned costs, it is possible to increase robustness of the proposed solution providing a good alternative to the solutions given by existing conflict-free trajectory planning models.

Keywords 4D trajectory management • Robust trajectory planning • Homotopy route generation model

A. Vidosavljevic (✉) • D. Delahaye
MAIAA-Applied Mathematics Laboratory, Ecole Nationale de l'Aviation Civile (ENAC), Toulouse, France
e-mail: vidosavljevic@recherche.enac.fr; delahaye@recherche.enac.fr

V. Tosic
Division of Airports and Air Traffic Safety, Faculty of Transport and Traffic Engineering, University of Belgrade, Beograd, Serbia
e-mail: v.tosic@sf.bg.ac.rs

Electronic Navigation Research Institute, *Air Traffic Management and Systems II*, Lecture Notes in Electrical Engineering 420, DOI 10.1007/978-4-431-56423-2_4

1 Introduction

Even at the current level of traffic demand, en route congestion is cited as one of the principal restricting factors to future growth of the airline industry. In 2012, about 17% (or 1.5 million) of flights in Europe arrived with more than a 15 min of delay compared with the schedule [24]. For the same year, IATA [15] has estimated that delays increased airline direct operating cost by 4.5 billion euros. With global air traffic demand expected to continue growing (predicted to be tripled by 2050 [30]), ICAO foreseen fundamental change in the operating paradigm for air navigation services in the following decade [16]. Major system development programs are underway around the world, including the Next Generation Air Transportation System (NextGen) program in the United States [23] and the Single European Sky ATM Research (SESAR) program in Europe [28].

Trajectory-based operation (TBO) represents a cornerstone of future ATM. TBO will enable aircraft to fly a negotiated flight path, termed reference business trajectory (RBT), taking into consideration both operator preferences (more direct and fuel-efficient routes) and optimal airspace system performance. Due to the high precision of RBT, TBO implies the possibility to design efficient congestion-free aircraft trajectories more in advance (pre-tactical, strategic level). Still, there are many uncertainty sources that may deflect an aircraft from its intended position (initial delay, wind, atmospheric temperature, actual aircraft mass, etc.) in addition to special events such as severe weather, volcanic ash, ATC strikes, etc. The best planning algorithm, however, is useless if the resulting plans cannot be implemented in the real world.

Therefore, in this work, model for robust 4D trajectory planning is proposed, taking into account uncertainty of trajectory prediction and unforeseen event while minimizing total additional cost incurred to the airspace users due to deviation from the user-preferred trajectories (UPT).

This paper is organized as follows. Section 2 contains a brief overview of the existing methods for trajectory planning. In Sect. 3, a new method for robust trajectory planning (RTP) is proposed including homotopy route generation model. The proposed RTP model is applied on a large-scale real-life problem, and test results are presented in Sect. 4. Finally, Sect. 5 sums up the findings, lists the major contributions of this work, and points toward areas of future research.

2 Previous Research

In recent years, problem of trajectory planning became increasingly popular. Several classes of methods are used to address this problem. One of the earliest approaches [3] is based on the evolution theory and uses basic operators, selection, mutation, and crossover, for generating a new population of conflict-free aircraft trajectories. The state space is a set of finite horizontal (straight line, turning point, and offset) and vertical (level-off) maneuvers, similar to the ones used currently by ATCo. To get closer to the real ATC system, the same authors in their future work presented in [2] and [13] took into account speed uncertainty and real aircraft performance models. In more recent work [1], the authors propose a new framework that separates the trajectory prediction and conflict detection models from the solver module. Hence, for a given scenario, a 4D matrix, containing conflict information indexed by aircraft pairs and maneuver pairs, is computed using any simulator. This 4D matrix provides all the necessary data for the solver and enables the use and comparison of various algorithms on the same problem instances. Another approach for conflict resolution is proposed in [4], where a sequence of maneuvers is controlled by a particle swarm optimization (PSO) algorithm. Although these models generate feasible trajectories from the aircraft performance perspective and take maneuvers cost into account, unfortunately, they are not adapted to curved trajectories.

Another class of methods, based on a force field, enables automatic generation of conflict-free trajectories with a mathematical proof. Besides difficulties imposed by aircraft performance (speed limits or trajectory smoothness) [25], there are successful implementations of the force field methods, like [26, 33] and [27]. However, the major drawbacks of a force field method are the continuous aircraft maneuver one obtains in response to the changing force field [20] and the complete absence of any optimization.

In [25], a new methodology for trajectory planning, using B-splines, is presented. Tactical aircraft conflict resolution is formulated as an optimization problem whose decision variables are the spline control points. In this work, aircraft are represented using a kinematic model to solve conflicts on the level flight (2D), evidencing local aspects of the method. Moreover, uncertainty of trajectory prediction is not considered. Another trajectory dimension reduction technique, using piecewise linear interpolation, is presented in [5]. In this work, *strategic de-confliction* of aircraft trajectories is addressed with the objective to minimize the number of conflicts using route and slot allocations. Uncertainty in aircraft position is taken into account through *freedom margin*, which is slightly higher (6 NM) than the separation norm. The cost of the solution is not considered in these works. Trajectory design based on the wave front propagation principle, termed *light propagation algorithm* (LPA), is introduced in [12]. Propagation is discretized in space and time, and a branch-and-bound algorithm is used to compute smooth geodesic with static and dynamic obstacles. In order to deal with the conflict resolution problem, LPA controls sequentially aircraft trajectories by selecting aircraft according to some given priority rule. This however leads to unfairness between aircraft particularly for large problem instances involving many aircraft. This limitation is partly overcome using shifting time window that still does not guarantee global optimum. Further LPA extension presented in [11] includes longitudinal uncertainty of aircraft position. It is shown that such variant of the problem is very difficult to solve. Even with the use of uncertainty reduction techniques, like *required time of arrival*, algorithm failed to find a solution (there remains around 10% of conflicts).

Most existing trajectory planning models consider finding conflict-free trajectories that barely take into account uncertainty of trajectory prediction. It is shown in this work that in the case of traffic disturbances, it is better to provide robust solutions; otherwise, a newly generated congestion problem can be hard and costly to solve. In this work, however, we did not investigate the trade-off between the probability of such disturbances and the effect on the objective function, i.e., trade-off between risk and cost. Knowledge about disturbance probabilities and their effects may indicate when robustness is actually needed and what is the appropriate level of robustness required.

3 Robust Trajectory Planning

In following section, we first introduce the problem and propose a new method for robust trajectory planning followed by mathematical formulation.

3.1 Problem Statement

In this work, we present an alternative way to deal with uncertainty in aircraft position (inability to cope with RBT) and unplanned situations, through building a more robust flight plan at the pre-tactical or strategic level. Robustness includes both reducing the likelihood of disruption and increasing the number of options to recover easily from a disruption. This way, the flight plan itself becomes less affected by such disturbances, reducing thereby the need for tactical actions. Further, it positively affects the tactical controller workload alleviating traffic management and conflict resolution tasks, which is the primary target of future ATM [16].

The method proposed in this work aims at generating a set of robust 4D trajectories while minimizing total additional costs incurred to the airspace users due to deviation from the UPTs. Although adding robustness to flight plans could result in higher airline planned costs, disruption costs (cost of management actions taken to resolve conflict on the tactical level due to disruptions) are likely to be reduced due to reduced need for tactical interventions, hopefully leading to reduction in airline real operating costs.

The RTP problem is formulated as a multi-objective problem that addresses how to generate a set of 4D trajectories from origin to destination in an optimal manner. It addresses assignment of horizontal route shape (2D route), vertical profile, and slot of departure, in order to manage the two confronted objectives: maximizing total robustness and minimizing total planned operating costs. It aims at finding system-optimal (SO) solution, a solution that is optimal from the viewpoint of the system as a whole. RTP might be further constrained by ATS capacities (airport capacity) or may include no-fly zones (severe weather cells, restricted or prohibited zones, etc.) that flights should avoid.

3.2 Robustness and Flight Interaction

Let us first define what robustness is, as definition is usually problem dependent [21]. ATM is a safety critical system, whose main task is to provide a safe flow of air traffic before making it punctual and expeditious. Therefore, in this work, robustness is considered as "the ability of a system to resist to changes without adapting its initial stable configuration" [32].

Another question concerns disturbances that have to be taken into account and indicators that quantify the robustness [29]. There are many uncertainty sources that can deflect an aircraft from its intended position (initial delay, wind, atmospheric temperature, actual aircraft weight, etc.). The difference between aircraft actual position and planned position may happen in space and/or time. Taking into account advanced future avionics [14], the trajectory will conform to the flight plan in the spatial dimension, while the longitudinal position on the trajectory may be subject to deviations in the temporal dimension [7]. Time uncertainty of position is, therefore, the main disturbance considered in this work.

Various indicators may be used to characterize robustness in the context of trajectory planning. An alternative way of quantifying the solution robustness is to measure its vulnerability to environmental change. While robustness describes the strength of a solution, such a vulnerability indicator would measure its weakness [29]. In this work, flight interaction is chosen as a measure of solution vulnerability and as an indirect measure of solution robustness.

The interaction between two flights is defined as a situation where flights compete for the same point in 4D space at the planning level. Unlike a conflict which has a fixed separation norm [22], interaction takes into account the aircraft position uncertainty propagation (deviations from the RBT). When aircraft positions coincide in 4D space, it results in maximum interaction that decreases as the distance (in space and time) between them increases. In the presence of uncertainty, minimization of interaction between flights at the planning level decreases conflict probability and therefore increases solution robustness.

In this work, flight interaction is defined as an exponential function of time separation, i.e., difference between the times of arrival at the *conflicting points*. Conflicting points are a pair of route positions that are separated by less than the given 3D norm. Usually these are points of route intersection; however, they can also be closely separated points of nonintersecting routes. For a pair of conflicting point p_k and p_m, with time separation $TS_{p_k p_m}$, *interaction magnitude* is computed by (1):

$$I_{p_k p_m} = \begin{cases} e^{-\tau \cdot TS_{p_k p_m}}, & \text{if } TS_{p_k p_m} < TS^*. \\ 0, & \text{otherwise.} \end{cases} \tag{1}$$

where the user-defined parameter τ controls the steepness of the exponential function and TS^* bounds flight interaction. Both parameters might vary depending on the test scenario.

With curved flight routes, multiple conflicting points between a pair of flight trajectories might exist. All those points must be taken into account, because the number of conflicting points *quantifies interaction* between two flights. Finally, the interaction between two flights f_i and f_j is computed by taking into account both the magnitude of the interaction between conflicting points and their quantity as in (2). Parameter D in the equation represents conflicting points separation norm.

$$I_{f_i f_j} = \sum_{(p_k, p_m) \in P_{ij}} I_{p_k p_m},$$
$$P_{ij} = \left\{ (p_k, p_m) \mid p_k \in f_i \wedge p_m \in f_j \wedge \overline{p_k p_m} \leq D \right\} \tag{2}$$

With classical trajectory representation, an ordered list of position vectors (samples) in 4D, a brute force approach for detection of conflicting points involves a pairwise comparison of trajectory samples. However, this is a time-consuming process, not suitable for a problem that involves a large number of trajectory samples [9]. In this work, a grid-based scheme developed in [17] is used for position vector comparison. In this approach, the airspace is discretized using a four-dimension grid with each cell having a unique address. Every position vector is associated with the address of the cell in the grid at which it belongs, and its interaction against all other vectors is computed by considering only vectors belonging to its own cell and its surrounding cells.

3.3 Route Generation Model

Both NextGen and SESAR expect more direct, fuel-efficient routes in the future ATM, enabled by satellite positioning and advance navigation technologies. This will free airspace from the "old highways in the sky" [23]. There is only one shortest direct route (geodesic) between two points. However, it may be unavailable due to obstacles or traffic congestion, or it might not be optimal taking into account wind on the route. As a consequence, alternative routes have to be considered. Theoretically, there are an infinite number of alternatives to a curve-shaped trajectory. This makes the discrete choice models unsuitable, since the choice set is infinitely large. Therefore, a continuous choice working directly with curves should be modeled. Furthermore, curves are objects belonging to spaces with infinite dimensions, and in order to manipulate such objects, it is necessary to reduce the dimension of the search space [8].

In this work, homotopy is used as a dimension reduction technique. It has been observed that homotopic feature of continuous functions may be used to map easily large continuous spaces using only a small number of parameters. This feature is exploited in the route generation model to design the shape of the horizontal route (alternative horizontal route) that is matched with the vertical profile to produce a 3D route. Finally, 3D shapes are completed with the time dimension in order to create trajectories.

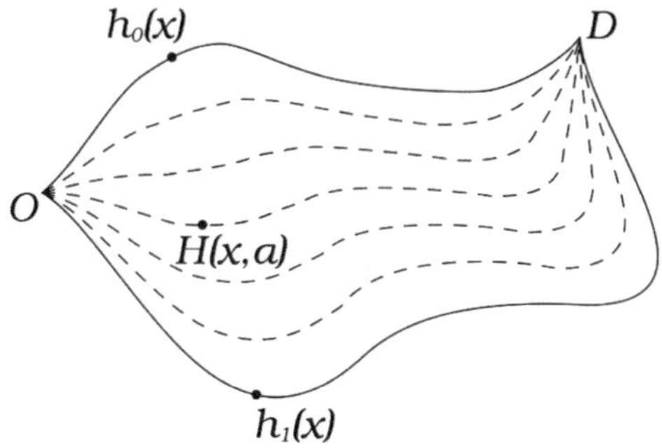

Fig. 1 Example of two homotopic curves in $\mathbb{R}^2$

3.3.1 Homotopy

In topology, two continuous functions are called homotopic if one can be "continuously deformed" into the other. Such a deformation is called a homotopy between the two functions, and the concept was first formulated by Poincaré around 1900 [6]. Formally, a homotopy between two continuous functions h_0 and h_1 from a topological space X to a topological space Y is defined as a continuous function $H : X \times [0, 1] \mapsto Y$ such that $H(x, 0) = h_0(x)$ and $H(x, 1) = h_1(x)$ for $\forall x \in X$ [19].

Dashed lines in Fig. 1 represent iso-contours of the homotopy $H(x, \alpha) = \alpha \cdot h_0(x) + (1 - \alpha) \cdot h_1(x)$ in $\mathbb{R}^2$ computed as a convex combination of the reference functions $h_0(x)$ and $h_1(x)$ for different values of the parameter α, $0 \leq \alpha \leq 1, \alpha \in \mathbb{R}$.

The route generation model uses a homotopy $H(x, \alpha)$ to map the space between reference functions using a single real-valued number α. Consequently, the model performance (resulting routes) is only influenced by the selection of the reference functions. Selection is further constrained as the route geometry[1] has a critical influence on the feasibility and performance of aircraft route tracking [26]. The list of reference function properties and selection process are presented in [31].

3.3.2 Mathematical Formulation

Figure 2 shows a pair of symmetric[2] reference functions $h_0 = h(x)$ and $h_1 = -h(x)$. For a given x, two points on the reference functions are represented as vectors in the Cartesian coordinate system by $\mathbf{p_1} = (x, h)$ and $\mathbf{p_2} = (x, -h)$, where $h = h(x)$. Hence, a point $\mathbf{p}$ of homotopy, defined as a convex combination with

[1] Routes have to be continuously differential, non-singular, etc.

[2] Symmetric reference functions have been selected so that the direct route, that is taken as nominal in this research, could be recovered.

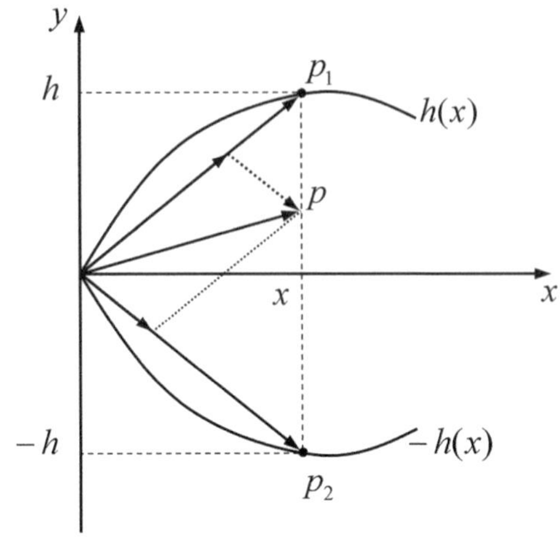

Fig. 2 Symmetric homotopy with respect to reference function $h(x)$

respect to the symmetric reference functions, is given by (3).

$$\begin{aligned} \mathbf{p} &= \alpha \cdot \mathbf{p_1} + (1-\alpha) \cdot \mathbf{p_2} \\ x_p &= \alpha \cdot x + (1-\alpha) \cdot x &&= x \\ y_p &= \alpha \cdot h + (1-\alpha) \cdot (-h) &&= (2\alpha - 1)\, h \\ \mathbf{p} &= \mathbf{p}(x, \alpha) &&= (x, (2\alpha - 1)\, h(x)) \end{aligned} \tag{3}$$

In the same manner, homotopy with respect to any symmetric reference functions $h^k(x)$ is given by $\mathbf{p^k}\left(x, \alpha^k\right)$ in formula (3). Finally, multiple homotopy, based on the reference functions $h^k(x)$, $k \in \{1, 2, \ldots, M\}$, is computed by (4), as a weighted arithmetic mean of corresponding homotopies. The weights $\left|\alpha^k - \frac{1}{2}\right|$ represent an absolute deviation of parameter α^k from $1/2$, which is considered "identity element" because the resulting symmetric homotopy is the direct (nominal) route.

$$\begin{aligned} \mathbf{p} = \mathbf{p}\left(x, \alpha^1, \alpha^2, \ldots, \alpha^M\right) &= \left(x, \sum_i Q^k \cdot h^k(x)\right) \\ Q^k = Q^k\left(\alpha^1, \alpha^2, \ldots, \alpha^M\right) &= \frac{2\left|\alpha^k - \frac{1}{2}\right| \cdot \left(\alpha^k - \frac{1}{2}\right)}{\sum_v \left|\alpha^v - \frac{1}{2}\right|} \end{aligned} \tag{4}$$

3.3.3 Alternative Route Design

Due to physical constraints on airports, it is assumed, in this research, that areas around airports and TMA will remain controlled. Therefore, the horizontal route shape is only modified in the en route segment. The route generation process is summarized in Fig. 3 and explained in details in the following paragraphs.

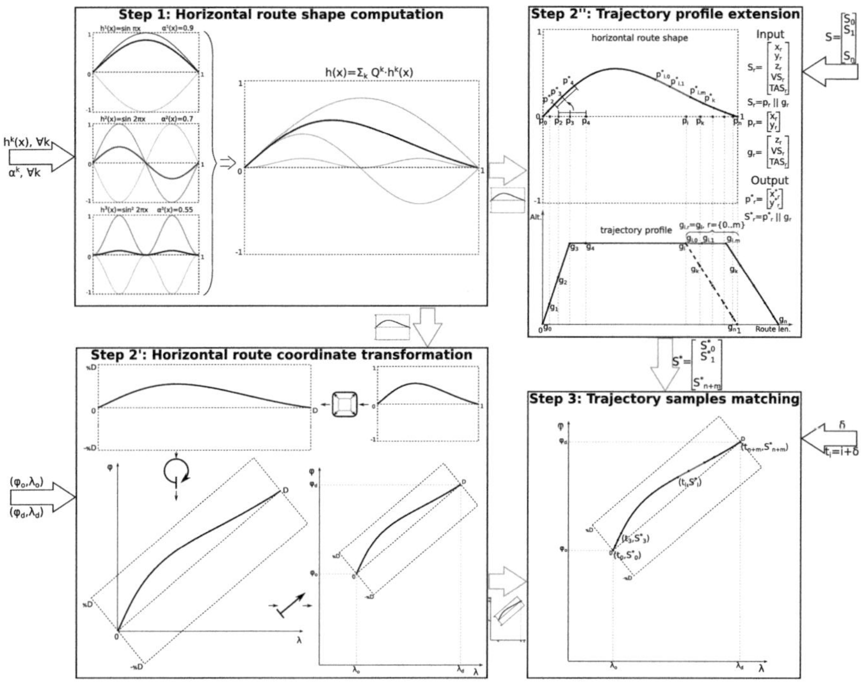

Fig. 3 Route generation model

In order to generalize the shape design process, making it independent of the origin-destination pair, the domain of the homotopy function is set to $[0, 1]$, and it takes a values in $[-1, 1]$, with the points $(0, 0)$ and $(0, 1)$ representing the start and end points of the en route segment. This also unifies route shape manipulations, like length calculation, trajectory profile extension, etc., for any possible flight. For a given set of homotopies with respect to symmetric reference functions $h^k(x)$ and set of parameters α^k that controls the homotopies, alternative horizontal route shape in en route segment is computed using the model presented in Sect. 3.3.2. The process is visualized in Fig. 3, Step 1. The horizontal route shape is decoded in Step 2 (Fig. 3) by scaling, rotation, and translation based on real coordinates of the start and end point.

The length of the alternative horizontal route is always larger than the length of the nominal direct route. Therefore, direct mapping between the given vertical profile and the new horizontal route is not possible. A solution is either to re-simulate the flight using the aircraft performance model, which is time consuming, or to approximate the vertical and speed profiles to match the new route length. In this work, the second approach is used, as it is shown in [5] that it is more efficient to extend the vertical profile at the top of descent in the cruise phase of the flight. The result is an acceptable approximation of the profile that respects optimal climb and descent gradients as well as the speed profile. Step 2" in Fig. 3 explains a process of trajectory profile extension. Based on a horizontal route extension and an initial trajectory state vector **S**, the trajectory profile is extended by adding flight segments as many times as required to make the length of the new profile and the length of the alternative horizontal route shape equal. The vectors $\mathbf{g}_{i.r}$ for $r = 0, 1, \ldots, m$, marked by a red line in the trajectory profile in Fig. 3, are copies of the original vector $\mathbf{g}_i$, for i being the top of descent point of the initial profile. The final trajectory is computed by matching samples of the new profile (vectors $\mathbf{g}_r$) with the new horizontal position vectors $\mathbf{p}^*{}_r$ for $r = 0, 1, \ldots, m + n$ completed with time.

3.4 RTP Mathematical Formulation

The following section describes mathematical modeling of the RTP, identifying decision variables, constraints, and objective functions, followed by mathematical program.

3.4.1 Nomenclature

In the mathematical formulation, the following list of nomenclature is used:

F	the set of flights f;
A_f	the set of indices of alternative vertical profiles for flight $f, f \in F$;
Δ	the set of possible ground delays;
M	the number of symmetrical homotopies;
a_{fj}	binary decision variable; equals to 1 if the flight f is assigned the alternative vertical profile j, and 0 otherwise, $f \in F, j \in A_f$;
δ_f	decision variable representing the (ground) delay of flight $f, f \in F$;
α_f^k	decision variable controlling homotopy k for flight f, $k \in \{1, 2, \ldots, M\}$, $f \in F$;
$t_f^{n.a}$	nominal arrival time of flight $f, f \in F$;
$t_f^{a.a}$	actual arrival time of flight $f, f \in F$;
$I_{f_1 f_2}$	flight interaction between flights f_1 and $f_2, f_1 \in F, f_2 \in F \backslash \{f_1\}$;
c_{fj}^p	initial cost of the alternative vertical profile j for flight $f, f \in F, j \in A_f$, that equals zero when a nominal vertical profile is used, $c_{fj_0}^p = 0$;
c_f^d	delay unit cost of flight $f, f \in F$;
c^g	fuel price per kilo;
FBR_{fj}	fuel burn rate for flight f using the vertical profile $j, f \in F, j \in A_f$ in [kg/min];
c^i	flight interaction unit cost;
C^p	total cost of alternative vertical profile;
C^d	total delay cost;
C^g	total fuel cost;
C^i	total flight interaction cost;

3.4.2 Decision Variables

The triple $(\delta_f, a_f, \alpha_f)$ represents a decision variable associated with each flight f in order to separate them in the 4D space. To separate trajectories in temporal space, a departure delay $\delta_f \in \Delta$ is assigned, while spatial separation is maintained by assignment of (a_f, α_f) for each flight in order to control their 3D routes. A vertical profile a_f is selected from the set of alternative vertical profiles, allowing change of cruising altitude. Vector $\alpha_f = \left(\alpha_f^1, \alpha_f^2, \ldots, \alpha_f^M\right)$ controls the shape of the horizontal route by controlling the homotopy H_f.

3.4.3 Constraints

In practice, flow control deals with a time interval divided into a finite number of periods, rather than with a continuous time variable. Therefore, departure delay can be treated as discrete by dividing the considered maximum allowed delay δ_{max} into v periods of equal length Δt ($v = \delta_{max}/\Delta t$). Then, the set of possible ground delays is $\Delta = \{0, \Delta t, \ldots, (v-1)\Delta t, v\Delta t\}$. An important feature of the homotopy route generation model is that the shape and length of an alternative horizontal route are bounded by the reference functions $h^k(x)$, $k \in \{1, 2, \ldots, M\}$. The only restriction is that the parameters controlling homotopies have to take a value between 0 and 1. Finally, the choice of vertical profile is limited to a predefined finite set of possible alternative profiles for each flight.

3.4.4 Objective Function

The RTP is formulated as an assignment of a vertical profile a_f, a vector of parameters α_f that controls homotopy $H_f(\alpha_f)$ and a delay δ_f for each flight f, such that the objective function consisting of the total additional costs to the network user and the total flight interaction cost is minimized. The total flight interaction is calculated as the sum of flight interactions between all pairs of flights, which, multiplied by unit interaction cost, gives total flight interaction cost (C^i). Total additional costs to network users, due to deviation from UPTs, are calculated as the sum of:

- Total cost of alternative vertical profile (C^p) – due to additional fuel burn and possible late arrival at destination. It is calculated as the sum of the initial cost c_{fj}^p, an input parameter associated with each alternative profile for all flights,
- Total delay cost (C^d) – due to late arrival at destination point. For each flight f, it is calculated as the difference between actual and nominal arrival times, where the nominal arrival time $t_f^{n.a}$ is an input parameter, while the actual time $t_f^{a.a}$ is computed by the route generation model. Delay unit costs and aircraft categorization are based on a study presented in [10],

- Total fuel cost (C^g) – due to longer alternative horizontal route. For each flight f, it is calculated based on the airborne delay and fuel burn rate FBR_{fj}, an input parameter that depends on the aircraft type of flight f and the cruising altitude of a profile j. *FBR* data were extracted from EUROCONTROL's Advanced Emission Model (AEM) that are based on BADA (Base of Aircraft Data) datasets.

3.4.5 Mathematical Model

$$\underset{(\delta,a,\alpha)}{\text{minimize}} \overbrace{\sum_{f\in F}\sum_{j\in A_f} c^p_{fj}\cdot a_{fj}}^{C^p} + \overbrace{\sum_{f\in F} c^d_f\cdot\left(t^{a.a}_f - t^{n.a}_f\right)}^{C^d}$$
$$+\overbrace{c^g\cdot\sum_{f\in F}\left[\left(t^{a.a}_f - t^{n.a}_f\right) - \delta_f\right]\cdot\sum_{j\in A_f} FBR_{fj}\cdot a_{fj}}^{C^g} \tag{5}$$
$$+\overbrace{c^i\cdot\sum_{f_1\in F}\sum_{f_2\in F\setminus\{f_1\}} I_{f_1f_2}}^{C^i}$$

subject to constraints:

$$\sum_{j\in A_f} a_{fj} = 1, \qquad \forall f\in F \tag{6}$$
$$\delta_f \in \Delta, \qquad \forall f\in F \tag{7}$$
$$0\le \alpha^k_f \le 1, \qquad \forall k\in\{1,2,\ldots,M\}, \qquad \forall f\in F \tag{8}$$
$$a_{fj}\in\{0,1\}, \qquad \forall j\in A_f, \qquad \forall f\in F \tag{9}$$
$$\alpha^k_f\in\mathbb{R}, \qquad \forall k\in\{1,2,\ldots,M\}, \qquad \forall f\in F \tag{10}$$

The first sum (C^p) in the objective function (5) represents the total initial cost due to the alternative vertical profiles. The second sum (C^d) is the total delay cost due to late arrivals at destinations. The third sum is the total cost of additional fuel burned due to longer routes. Finally, the last sum represents the total flight interaction cost.

Constraints (6) ensure that each flight can only be assigned to one vertical profile from the set of alternative vertical profiles. Constraints (7) stipulate that every flight delay takes a value from a given set of possible ground delays. Finally, constraints (8) indicate that the parameters controlling homotopies take a value between 0 and 1. Equations (9, 10) represent decision variable domain constraints.

4 Results and Discussion

The RTP problem, presented in the previous section, is known to be an NP-hard problem. In addition, this optimization problem is non-separable as the objective function is not separable, i.e., it cannot be expressed as the sum of the functions of the individual decision variables due to the links induced by flight interactions. Moreover the objective function is evaluated by simulation – black box evaluation. To solve real instances of problems involving high combinatories in a huge state space, one must rely to stochastic methods of optimization. Due to size of memory required for defining a point in the state space, simulated annealing (SA), local search metaheuristic, has been selected to address this problem based on the fact that it requires less memory than other techniques. More details on the size and complexity of the problem can be found in [31].

The RTP model together with an SA implementation is applied to a large-scale real-life example to test its capabilities against conventional conflict-free models.

4.1 Experimental Setup

This subsection describes the design of the experiment, traffic sample with two scenario settings.

4.1.1 Traffic Sample

The traffic data we consider includes traffic in the French metropolitan airspace on August 17, 2008, that consists of 8,845 flights. To make the problem size manageable, but still realistic, a 3-hour period (9:00–12:00) from the morning peak is selected. Finally, the traffic sample includes 1,755 flights, of which 50% are overflights, about 10% are domestic flights, and about 40% are international flights, having either departure or arrival at French airports. Flights are operated by 50 different aircraft types, which furthermore confirms the high heterogeneity of demand.

4.1.2 Traffic Data Source and Format

The system entry times (departure time or time at which flight enters airspace) and the system entry and exit points (airport or airborne fix) are extracted from real traffic data obtained from the French civil aviation system for flight plan processing, real-time radar data tracking and visualization – CAUTRA. Then, flights are simulated in the ENAC traffic simulator (CATS) [2] using direct flight routes. Resulting nominal trajectories with system exit times are then recorded. As CATS is a discrete model, the final flight data were provided as a sequence of route points sampled every 15 s. Each point contains the 4D position, the velocity, and the aircraft intention. Two successive points define a flight segment as a direct portion of flight.

4.1.3 Solution Space

Each flight is simulated using five different cruising altitudes, a nominal altitude and two lower and two higher cruising altitudes, maintaining an optimal vertical profile. At the end, each flight is assigned with five vertical profiles corresponding to these flight levels.

The shape of the alternative horizontal route is computed using the route generation model based on three reference functions, as explained in Fig. 3, Step 1, and three real parameters α^k with values between 0 and 1.

The maximum ground delay is set to 30 min.

4.1.4 Scenario Settings

In this work, two main scenarios are defined and tested. First is the base scenario, aiming to find a robust solution as a balance between interactions and additional flight costs, as defined in Sect. 3. The second scenario aims at finding a conflict-free solution not taking into account solution robustness. The second, supplementary scenario, is designed in order to evaluate the results of the base scenario.

For both scenarios, the same interaction, delay, and fuel unit costs are used, with difference in TS^*, that bounds flight interaction. Interaction is bound to 3 min in the base scenario, while it is set to zero for conflict-free scenario, meaning that it is sufficient for a flight to be separated in space or in time.

4.2 Experimental Results

Scenarios are tested on an Intel Pentium Dual-Core 2.9 GHz PC with 4 GB of RAM.

4.2.1 Base Scenario

When robustness is taken into account, the number of conflicting points of the initial flight plan (nominal trajectories) is 310,814. It includes flight interactions of different magnitude resulting in 92.66 million euros of total initial interaction cost, as an initial value of the objective function. This cost penalizes flight interactions existing in the initial flight plan and is not related to the real airspace user costs.

In the search for the best robust solution, the optimization algorithm is able to reduce the value of the objective function to 0.19 million euros. The algorithm intended to find the best possible balance between interaction and operating costs; hence, there remains interactions located near Paris TMA (marked in red in Fig. 4a). This is, however, a very challenging task. The best solution is obtained in 20 h of CPU time, while the solution with objective value in the range of 10% of the objective value of the best-known solution is found in less than 500 SA steps, in 13 h. For the sake of comparison, the variant of the problem excluding flight operating costs from the objective function yields a robust solution in 1 h on average.

To find a robust solution, as expected, many flights are modified: 87% of all flights. The majority of flights are assigned an alternative horizontal route, while less than 30% are delayed or assigned to a non-nominal cruising altitude. The main numerical results are summarized in Table 1.

Required trajectory modifications result in an increase in operating cost of 108 euros per flight on average. However, the operating cost increase is not

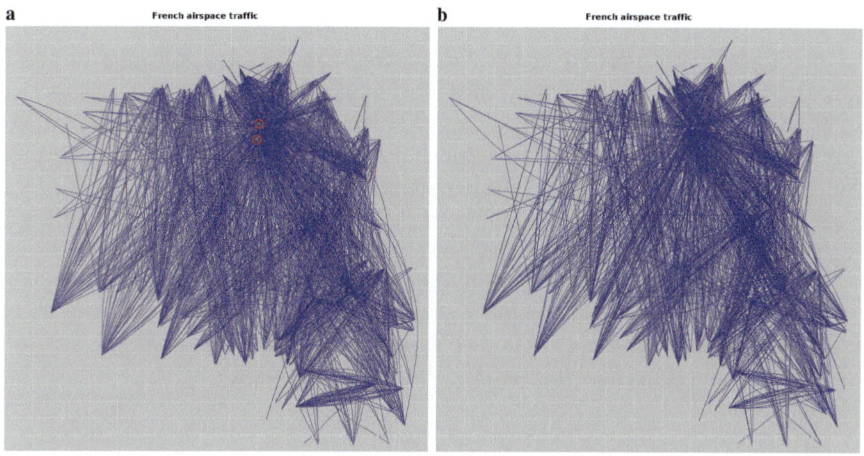

Fig. 4 Resulting solution trajectories: (**a**) Robust scenario. (**b**) Conflict-free scenario

Table 1 Experimental results

	Conflict-free	Base
Modified flights	**45%**	**87%**
Horizontal route	40%	85%
Ground delay	15%	30%
Vertical profile	2%	23%
Total route extension	**0.3%**	**0.8%**
Avg. per extended flight	0.6%	
Total en route delay	250 min	781 min
Total ground delay	**342 min**	**2743 min**
Maximum ground delay	22 min	27 min
Avg. per delayed flight	1.4 min	5 min
Avg. flight level change	**1.2 fl**	**1.3 fl**
Total additional operating costs	**29,984 €**	**192,417 €**

evenly distributed across airspace users. In addition to the average value, as a first approximation of the solution quality, it is reported that 90% of flights have an increase in operating costs lower than 300 euros. Maximum cost increase recorded is 1,288 euros. Although it is expected that costs of system-optimal solution is shared among airspace users [18], the authors are aware of the equity issue that has to be taken into account in further research. Nevertheless, this is a very promising result considering that no additional disruptions should be experienced due to the obtained solution's increased robustness.

4.2.2 Conflict-Free Scenario

Considering only the conflicts, the general problem is then relaxed, and a solution to the problem is easier to find. The initial number of conflicting points between nominal flight trajectories is 18,268, resulting in total initial cost of 18.27 million euros.

The algorithm is able to find a conflict-free solution (Fig. 4b), and the best solution found has value of the objective function equal to 29,984 euros. The solution is obtained in 16 h of CPU time, while the solution with the objective value in the range of 10% of the objective value of the best-known solution is found in less than 200 SA steps in 7 h. For the sake of comparison with other existing conflict-free trajectory planning models, the variant of the problem excluding flight operating costs from the objective function may be solved in less than 30 s.

In the solution found, more than half of the trajectories are not modified, and once again a horizontal route modification is mostly used to solve potential conflicts. Due to alternative trajectories, operating cost increased by 17 euros per flight, on average, with 90% of flights having costs increase lower than 50 euros.

4.3 Solution Robustness Testing

To test solution robustness, conflict-free (non-robust) and robust (base) scenario solutions are imposed to additional flight delays, and effects of these disruptions are measured by comparing congestion problems they produce to the system. The magnitude of the congestion problem is measured by the number of newly generated conflicts.

Several scenarios are defined varying in disruption level, as shown in Table 2, depending on the number of affected flights. The delay of affected flights is fixed to 1.5 min for all test scenarios. It is shown that larger delay does not necessarily lead to larger disruption and that disruption level is more influenced by the number of delayed flights. Furthermore, intensive computational experiments revealed that the resulting congestion problem is very sensitive to which specific flights are chosen (by the algorithm) to be delayed. As a consequence, multiple scenario repetitions are performed for each test.

Table 2 Robustness test scenario settings

Scenario	Number of affected flights	Delay
1	20	1.5 min
2	50	1.5 min
3	100	1.5 min
4	500	1.5 min
5 (chaos)	1000	1.5 min

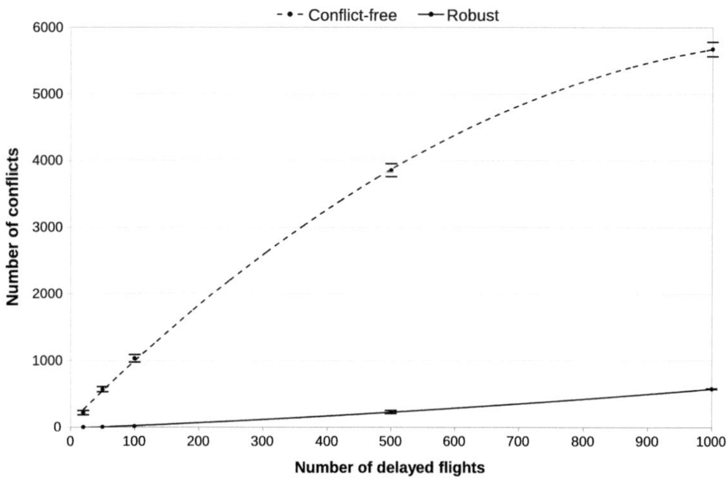

Fig. 5 Mean, trend of the mean, and 95% confidence interval of the number of newly generated conflicts

The effect of flight plan disruptions on the number of newly generated conflicts is illustrated in Fig. 5. The disruption level is represented by the number of delayed flights. The figure shows the mean value of the number of conflicts for five robustness test scenarios (Table 2), 95% confidence interval on the mean, and the interpolation of the mean number of conflicts. It clearly demonstrates that robust trajectories are less affected by disruptions and therefore cause fewer congestion problems. Interpolation reveals polynomial trend between the number of delayed flights and the number of conflicts for conflict-free scenario, while their correlation is almost linear for robust scenario. Moreover, the size of confidence interval increases with the disruption increase for the conflict-free scenario indicating it is less stable than robust scenario whose confidence interval is not changed with disruption level. Finally, it is shown, during the line of this research, that it is easier to solve congestion problems of robust trajectories at the tactical level, compared to the one of (non-robust) conflict-free trajectories.

5 Conclusions

This work investigates the potential of robust trajectory planning at the strategic and pre-tactical levels as a means to alleviate the en route congestion in airspace. It introduces a novel approach for route shape generation based on homotopic feature of continuous functions. This approach is capable of generating a large number of routes of different shape with a reasonable number of decision variables.

Application of the proposed optimization model and algorithm is illustrated in a real-life example. Results show that the model is able to solve real instances of the problem, within reasonable computation (computation time corresponds to the intended use of the model).

Further, the results indicate that, under certain conditions, solution robustness could be considerably increased at the relative small expense of the solution cost, providing a good alternative to the solutions developed by existing conflict-free trajectory planning models.

Introduction of operating costs into the objective function significantly increases the problem complexity, as, due to the rugged shape of the objective function, slow exploration of the search space is inevitable. This influences computation time that should be further improved in order to address larger problem instances. Furthermore, a feature of the homotopy route generation model, the fact that the horizontal route is controlled by real-valued parameters, could be more exploited in future research using field congestion metrics. It is foreseen that in such a case, it would be possible to represent congestion metrics as an explicit function of such parameters that controls each homotopy, instead of the indirect computation of the objective function as we did in this paper throughout the simulation.

Acknowledgements The research of this work has been supported by research grants of the Ministry of Science and Technological Development, Republic of Serbia, through projects of the Faculty of Transport and Traffic Engineering funded under the 2011–2014 research programs in technological development: Project TR36033, "A support to sustainable development of the Republic of Serbia's air transport system."

The authors would like to thank Supatcha Chaimatanan whose programming code is used as a basis for the development of the optimization model, the French Civil Aviation University (ENAC) for providing traffic data, and especially Cyril Allignol for his assistance in CATS data manipulation. The authors are also immensely grateful to Professor Marcel Mongeau (ENAC) for the review of the manuscript and his valuable comments.

References

1. Allignol, C., Barnier, N., Durand, N., Alliot, J.M.: A new framework for solving en-route conflicts. In: Proceedings of 10th USA/Europe Air Traffic Management Research and Development Seminar (ATM2013). Chicago, IL (2013)
2. Alliot, J.M., Durand, N.: Optimal resolution of en route conflicts. Tech. rep., CENA, Toulouse, France (1997)

3. Alliot, J.M., Gruber, H., Joly, G., Schoenauer, M.: Genetic Algorithms for solving Air Traffic Control conflicts. Tech. rep., CENA, Toulouse, France (1993)
4. Blasi, L., Barbato, S., Mattei, M.: A particle swarm approach for flight path optimization in a constrained environment. Aerospace Science and Technology **26**(1), 128–137 (2013)
5. Chaimatanan, S., Delahaye, D., Mongeau, M.: Hybrid metaheuristic optimization algorithm for strategic planning of 4D aircraft trajectories at the continent scale. IEEE Computational Intelligence Magazine **9**(4), 46–61 (2014)
6. Collins, G.P.: The shapes of space. Scientific American **291**, 94–103 (2004)
7. Delahaye, D., Puechmorel, S.: Air traffic complexity based on dynamical systems. In: Proceedings of 49th IEEE Conference on Decision and Control (CDC2010). Atlanta, GA (2010)
8. Delahaye, D., Puechmorel, S.: Modeling and Optimization of Air Traffic. Wiley-ISTE (2013)
9. Delahaye, D., Puechmorel, S., Tsiotras, P., Feron, E.: Mathematical models for aircraft trajectory design: A survey. In: Proceedings of 3rd ENRI International Workshop on ATM/CNS (EIWAC2013). Tokyo, Japan (2013)
10. Department of Transport Studies - University of Westminster: European airline delay cost reference values. Tech. rep., University of Westminster for PRU EUROCONTROL, London, UK (2011)
11. Dougui, N., Delahaye, D., Mongeau, M., Puechmorel, S.: Aircraft trajectory planning under uncertainty by light propagation. Procedia - Social and Behavioral Sciences **54**, 201–210 (2012)
12. Dougui, N., Delahaye, D., Puechmorel, S., Mongeau, M.: A light-propagation model for aircraft trajectory planning. Journal of Global Optimization **56**(3), 873–895 (2013)
13. Durand, N., Alliot, J.M., Noailles, J.: Automatic aircraft conflict resolution using Genetic Algorithms. In: Proceedings of the Symposium on Applied Computing (SAC'96). Philadelphia, PA (1996)
14. Episode 3 project consortium: Overall description of the platform and its capabilities. Tech. rep., EUROCONTROL, Brussels, Belgium (2009)
15. IATA Economics: IATA economic briefing - Inefficiency in european airspace. Tech. rep., International Air Transport Association, Montreal, Quebec, Canada (2013)
16. ICAO: Global Air Traffic Management operational concept, Doc 9854, First Edition. Tech. rep., International Civil Aviation Organization, Montreal, Quebec, Canada (2005)
17. Jardin, M.: Real-time conflict-free trajectory optimization. In: Proceedings of 5th USA/Europe Air Traffic Management Research and Development Seminar (ATM2003). Budapest, Hungary (2003)
18. Jovanovic, R.: Economic measures for air traffic demand management. Ph.D. thesis, UB-FTTE, University of Belgrade - Faculty of Transport and Traffic Engineering, Belgrade (2011)
19. Katok, A., Sossinsky, A.: Lecture notes: Introduction to modern topology and geometry. http://www.personal.psu.edu/axk29/MASS-07/Background-forMASS.pdf (2006)
20. Kuchar, J., Yang, L.: A review of conflict detection and resolution modeling methods. IEEE Transactions in Intelligent Transportation Systems **1**(4), 179–189 (2000)
21. Lan, S.: Planning for robust airline operation: Optimize aircraft routings and flight departure times to achieve minimum passenger disruptions. Tech. rep., Massachusetts Institute of Technology, Cambridge, MA (2003)
22. Netjasov, F., Vidosavljevic, A., Tosic, V., Everdij, M., Blom, H.: Development, validation and application of stochastically and dynamically coloured Petri net model of ACAS operations for safety assessment purposes. Transportation Research Part C **33**, 167–195 (2013)
23. NextGen: NextGen - Implementation Plan. Tech. rep., US Department of Transportation, FAA, Washington, D.C. (2012)
24. Performance Review Commission - PRC: PRR 2012 - Performance Review Report, An assessment of Air Traffic Management in Europe during the calendar year 2012. Tech. rep., EUROCONTROL, Brussels, Belgium (2013)

25. Peyronne, C., Conn, A.R., Mongeau, M., Delahaye, D.: Solving air-traffic conflict problems via local continuous optimization. European Journal of Operational Research **241**(2), 502–512 (2015)
26. Roussos, G., Chaloulos, G., Kyriakopoulos, K., Lygeros, J.: Control of multiple non-holonomic air vehicles under wind uncertainty using Model Predictive Control and decentralized navigation functions. In: Proceedings of 47th IEEE Conference on Decision and Control (CDC'08). Cancun, Mexico (2008)
27. Roussos, G., Dimarogonas, D., Kyriakopoulos, K.: 3D navigation and collision avoidance for nonholonomic aircraft-like vehicles. International Journal of Adaptive Control and Signal Processing **24**(10), 900–920 (2010)
28. SESAR consortium: SESAR Defenition Phase: The concept of operations at a glance. Tech. rep., EUROCONTROL, Brussels, Belgium (2013)
29. Snelder, M., van Zuylen, H., Immers, L.: A framework for robustness analysis of road networks for short term variations in supply. Transportation Research Part A **46**, 828–842 (2012)
30. STATFOR, the EUROCONTROL Statistics and Forecast Service: Challenges of growth 2013 - European air traffic in 2050. Tech. rep., EUROCONTROL, Brussels, Belgium (2013)
31. Vidosavljevic, A.: Trajectory planning on pre-tactical and tactical level in Air Traffic Management. Ph.D. thesis, UB-FTTE, University of Belgrade - Faculty of Transport and Traffic Engineering, Belgrade (2014)
32. Wieland, A., Wallenburg, C.M.: Dealing with supply chain risks - linking risk management practices and strategies to performance. International Journal of Physical Distribution & Logistics Management **42**(10), 887–905 (2012)
33. Zeghal, K.: A review of different approaches based on force field for airborne conflict resolution. In: Proceedings of AIAA - Guidance, Navigation and Control Conference. Boston, MA (1998)

Numerical Investigation on Flight Trajectory Optimization Methods

Akinori Harada

Abstract This paper evaluates the usability of two direct optimization methods: the Piecewise Linear Approximation Dynamic Programming method and the gradient-based method for a practical trajectory optimization problem based on the trajectory-based operation concept. As a practical application, the longitudinal fuel minimal trajectory of a passenger aircraft is continuously designed for all flight phases. The computational features of each method are investigated in terms of computational time, convergence to a realistic solution, and applicability for the practical model. The two kinds of optimal trajectories show a great agreement indicating that an optimal solution close to a global optimum could be obtained by the gradient-based method. The results have demonstrated that gradient-based methods have a potential to provide an optimal solution close to a global optimum with a reasonable computational time. The gradient-based method is expected to be utilized for designing a practically preferable reference trajectory toward the realization of more efficient operations in the air traffic management system.

Keywords ATM • Optimal control • Trajectory optimization • Dynamic programming • Gradient-based method

Nomenclature, Subscripts, Superscripts

Nomenclature

C	Constraints, coefficient
D	Drag
F	Constraint vector
f	Function vector
g	Gravity acceleration

A. Harada (✉)
School of Systems Engineering, Kochi University of Technology, Kochi, 782-8502, Japan

Graduate School of Engineering, Kyushu University, Fukuoka, 819-0395, Japan
e-mail: harada.akinori@kochi-tech.ac.jp

Electronic Navigation Research Institute, *Air Traffic Management and Systems II*, Lecture Notes in Electrical Engineering 420, DOI 10.1007/978-4-431-56423-2_5

H	Hamiltonian, altitude
J	Objective function
k	Induced drag coefficient
L	Lagrangian, lift
M	Mach number
m	Dimension of the control input, mass of aircraft
n	Dimension of the state variable, number of elements
T	Thrust
δT	Thrust lever position
t	Time
u	Control input
V	True airspeed
X	Downrange
x	State variable
y	Output variable
Z	Optimization variable vector
γ	Flight-path climb angle
η	Cross range angle
θ	Longitude
μ	Fuel flow
ξ	Downrange angle
ρ	Air density
ϕ	Latitude, objective function at terminal
ψ	Constraint function at terminal, flight-path heading angle

Subscripts

$0, f$	Initial, final
$D0$	Parasite drag
i, j	Element numbers for vector and matrix
k	Time stage number
L	Lower
max	Maximum
min	Minimum
MO	Maximum operating
opt	Optimal
s	Shooting
U	Upper

Superscripts

o	Optimal
*	Stationary point

1 Introduction

Over the past decades, world air traffic has been growing by the global increase of population and improvement of economic standard. Recently, air traffic is growing continuously and is anticipated to be doubled within the next fifteen years. Rising fuel consumption and emissions and the flight safety deterioration caused by the increasing air traffic are urgent issues. More efficient operations have become possible by the current highly developed CNS/ATM technologies. Under the NextGen program in the United States and SESAR in Europe, many research projects are implemented to realize a promising ATM system. In Japan, Collaborative Actions for Renovation of Air Traffic System (CARATS) has been conducted by the government as a roadmap for developing Japanese future air transportation system. One of the significant plans for the future ATM system is an innovative change from existing airspace-based operations to the trajectory-based operations (TBO) which deal with the flight path from departure to arrival as one continuous trajectory on the unified airspace. The TBO is foreseen to become a key component to realize more efficient operations on the way toward the ideal ATM system. Although flight trajectory optimization is an essential technique for the realization of TBO, few research have been carried out for developing optimization tools which can provide practically preferable reference trajectories for all flight phases.

As a practical trajectory optimization method, Dynamic Programming (DP) has been mainly used to evaluate the potential benefits of CARATS [1–7]. DP has many powerful advantages such as global optimality, no iterations for the convergence, and easiness for handling inequality constraints. The well-known disadvantage, however, the so-called curse of dimensionality, still remains a difficult challenge even with the high-processing capacity of recent computers. In the optimal control field, gradient-based algorithms available in generalized optimization tools have been widely applied as a versatile optimization method. This method has capabilities for solving high-dimensional problems such as multiple aircraft trajectory optimizations also considering conflicts [8]. This paper designs fuel-efficient optimal trajectories for all flight phases as a practical and significant application in considering realization of TBO by both DP and the gradient-based method. The base of aircraft data (BADA) family 3 is applied as the performance model. An accuracy evaluation with flight data obtained from aircraft on-board system has shown that the error of BADA model is at a reasonable level for designing a realistic optimal trajectory [9].

The main purpose of this paper is to evaluate the usefulness of each method for the trajectory optimization task by revealing their merits in terms of convergence to a realistic solution, computational time and applicability for the practical model.

This paper consists of three parts. The general optimal control problem is stated in Chap. 2. The basic algorithms of Piecewise Linear Approximation Dynamic Programming (PLA-DP) method and the gradient-based method are introduced with an emphasis on promising features in Chap. 3. In Chap. 4, the fuel minimal trajectories are designed by the two methods, and finally, those trajectories are evaluated with regard to the computational features to investigate those methods' usability in practical trajectory optimization tasks.

2 Statement of Optimal Control Problem

The general optimal control problem (OCP) for continuous-time systems can be described as determining the optimal control histories

$$\boldsymbol{u}_{opt}(t) \in \mathbb{R}^{m} \tag{1}$$

and the corresponding optimal trajectories

$$\boldsymbol{x}_{opt}(t) \in \mathbb{R}^{n} \tag{2}$$

which minimize the objective function

$$J = \int_{t_0}^{t_f} L(\boldsymbol{x}(t), \boldsymbol{u}(t), t) dt + \phi\left(\boldsymbol{x}\left(t_f\right), t_f\right) \tag{3}$$

subject to the state equation

$$\dot{\boldsymbol{x}} = \boldsymbol{f}(\boldsymbol{x}, \boldsymbol{u}) \tag{4}$$

and the equality and inequality constraints

$$\boldsymbol{C}_{eq}(\boldsymbol{x}(t), \boldsymbol{u}(t), t) = 0 \tag{5}$$

$$\boldsymbol{C}_{ineq}(\boldsymbol{x}(t), \boldsymbol{u}(t), t) \leq 0 \tag{6}$$

as well as the initial and final boundary conditions.

$$\psi_0(\boldsymbol{x}(t_0), \boldsymbol{u}(t_0), t_0) = 0 \tag{7}$$

$$\psi_f\left(\boldsymbol{x}\left(t_f\right), \boldsymbol{u}\left(t_f\right), t_f\right) = 0 \tag{8}$$

Additionally, the output variable vector is obtained.

$$y(t) = Cx(t) \tag{9}$$

3 Flight Trajectory Optimization Methods

Techniques for solving optimal control problems can be classified roughly into either a direct method or an indirect method. As an indirect method, the calculus of variations where Euler-Lagrange differential equations are solved as a two-point boundary-value problem has been established by early studies. In the indirect method, the range of applications is restricted due to the following difficulties as Betts mentions in reference [10].

Difficulties in an Indirect Method

1. The knowledge of optimal control theory and special skills are required to compute the quantities that appear in defining adjoint equations, control equations, and transversality equations.
2. It is extremely difficult to make an a priori estimate of the constrained-arc sequence if the problem description includes path inequalities.
3. The user must guess values for the adjoint variables which are not physical quantities. The numerical solution of the adjoint equations can be ill conditioned and sensitive. The solution inevitably causes a lack of robustness.

These difficulties prevent an indirect method from being extensively used in practice. Meanwhile, a direct method, in which a finite number of parameters indicate the trajectory and are solved as a parameter optimization problem, has become increasingly used, thanks to the improvements in the processing capability and memory capacity of recent computers. A direct method does not need any complicated formulation even if the problem includes multiple inequality constraints. Once the optimal control problem is defined appropriately, the calculation can be performed easily using a variety of numerical optimization libraries available in several programming languages. This section introduces two methods, the Piecewise Linear Approximation Dynamic Programming (PLA-DP) method and the gradient-based method as the promising approaches in a direct method. How to apply these methods to a general trajectory optimization problem is explained with the fundamental concepts.

3.1 PLA-DP Method

Dynamic Programming (DP), firstly proposed by Richard E. Bellman in the 1950s [11], is an optimization theory based on the Hamilton-Jacobi-Bellman (HJB) first-order nonlinear partial differential equation denoted by Eq. (10) [12].

$$-\frac{\partial J_{opt}}{\partial t} = \min_{\boldsymbol{u}} H\left(\boldsymbol{x}, \frac{\partial J_{opt}}{\partial \boldsymbol{x}}, \boldsymbol{u}, t\right) \tag{10}$$

DP describes the system by finite quantized grid points and solves the combinatorial optimization problem and, thereby, is classified as a direct method. DP has many fascinating advantages over other methods.

Advantages in DP

1. Global optimality
   DP defines a grid in the search space of the state variables and finds the optimal solution based on the HJB optimality conditions. Therefore, the algorithm assures the global optimum, not a local optimum that satisfies only the necessary conditions.
2. Predictable computational time
   The computational process for DP is decisive and does not contain any iterative calculations for convergence. Hence, the computational time may be estimated in advance because it depends only on the number of grid points and the number of state variables in the state space.
3. Easiness in handling inequality constraints
   Inequality constraints may easily be included for the state variables themselves and their functions. Rather, inequality constraints are suitable for limiting the search space. Inequality constraints on control inputs can be also implemented easily.
4. Simplicity for coding
   The discrete form of the HJB optimality condition makes the programming code for this method comparatively simple and easy to be understood.

In contrast, DP has the following three major disadvantages which generally limit its practical usage in engineering applications.

Disadvantages in DP

1. Curse of dimensionality [13, 14]
   Computational time and memory usage increase explosively with the number of state variables and the number of divisions.
2. Menace of the expanding grid [15]
   If the number of control inputs m is the same as the number of state variables n, DP calculation may be easily performed. In case of $m < n$, the solution cannot be derived in the state transition calculation between two quantized grid points.
3. Quantization error
   The state and the control inputs need to be quantized using a predefined grid. The states may not take any continuous value but only values from the predefined set.

The Moving Search Space Dynamic Programming (MS-DP) which was proposed by Miyazawa [16] has the capability to alleviate the first disadvantage and has discovered a potential to enable a real-time trajectory optimization.

To relieve the second disadvantage, two relaxation methods, the least-error grid point selection (LGS) method and the augmented control variable (ACV) method, were proposed maintaining the combinatorial optimization concept in the previous literature [17, 18]. Another approach based on the interpolation concept was devised by Lapidus and Luus [19] and Matsuda [20] and completed as PLA-DP method [20, 21]. The algorithm of PLA-DP is briefly introduced using a simple control system denoted by Eqs. (11, 12, 13).

$$\dot{\boldsymbol{x}} = \boldsymbol{f}(\boldsymbol{x}(t), \boldsymbol{u}(t), t),\ \boldsymbol{x} \in \mathbb{R}^2,\quad \boldsymbol{u} \in \mathbb{R}^1 \tag{11}$$

$$\boldsymbol{x}(t) = \begin{pmatrix} x_1(t) \\ x_2(t) \end{pmatrix} \tag{12}$$

$$\boldsymbol{u}(t) \in u_1(t) \tag{13}$$

The duration $t \in [t_0, t_f]$ is discretized into $n_{\Delta t} + 1$ time points.

$$t_k = t_0 + \left(t_f - t_0\right)(k-1)/n_{\Delta t},\quad k = 1, 2, \cdots, n_{\Delta t} + 1 \tag{14}$$

$n_{\Delta t}$: Number of time fractions in the duration

Each state variable is quantized into $n_i \quad (i = 1, 2)$. The j th value of x_i is written as:

$$\begin{aligned} x_i(j) &= x_{i,\min} + (x_{i,\max} - x_{i,\min})(j-1)/n_i, \\ j &= 1, 2, \cdots, n_i + 1, x_{i,\min} \le x_i(j) \le x_{i,\max} \end{aligned} \tag{15}$$

The optimality condition for this system arises from the *Principle of Optimality* [11] denoted by Eq. (16).

$$J_{opt_k}\left(x_1(j_1), x_2(j_2), t_k\right) = \min_{u_1}\left[L\Delta t + J_{opt_{k+1}}\left(x_1\left(j_1'\right), x_2, t_{k+1}\right)\right] \tag{16}$$

If the control input is set by zero-order hold over the transition between the kth stage and $k+1$th stage, the feasible transition point from the grid point at the kth stage does not hit on any grid points at the $k+1$th stage. The feasible transition points may be determined analytically on a blue line as illustrated in Fig. 1.

The optimum return function (ORF) value at the $k+1$th stage, $J_{opt_{k+1}}$ which is necessary to execute the optimization calculation presented by Eq. (16) recursively, is derived by the following process.

Approximation of the ORF Value

1. The ORF value is already stored on all the grid points at the $k+1$th stage by the former calculation.

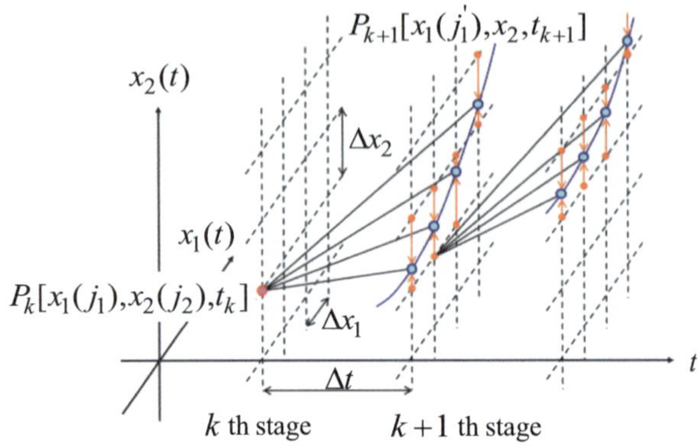

Fig. 1 Approximation of the ORF Value

2. The feasible points $P_{k+1}\left[x_1\left(j_1^{'}\right), x_2, t_{k+1}\right]$ at the $k+1$th stage are obtained analytically for all grid point values of $x_1\left(j_1^{'}\right)$.
3. The ORF value at the point P_{k+1} is calculated by linear interpolation using two neighboring grid points along the x_2 axis. The following equation explains the interpolation for the two grid points $x_2\left(j_2^{'}\right)$ and$x_2\left(j_2^{'}+1\right)$:

$$
\begin{aligned}
&J_{opt_{k+1}}\left(x_1\left(j_1^{'}\right), x_2, t_{k+1}\right) = J_{opt_{k+1}}\left(x_1\left(j_1^{'}\right), x_2\left(j_2^{'}\right), t_{k+1}\right) \\
&+\left[J_{opt_{k+1}}\left(x_1\left(j_1^{'}\right), x_2\left(j_2^{'}+1\right), t_{k+1}\right) - J_{opt_{k+1}}\left(x_1\left(j_1^{'}\right), x_2\left(j_2^{'}\right), t_{k+1}\right)\right]\cdot \\
&\left[x_2 - x_2\left(j_2^{'}\right)\right] / \left[x_2\left(j_2^{'}+1\right) - x_2\left(j_2^{'}\right)\right]
\end{aligned}
\tag{17}
$$

   This interpolation for two grid points is illustrated by the orange arrows in Fig. 1.

4. The fraction of the objective function $L\Delta t$ denoted in Eq. (16) is calculated between P_k and the feasible points P_{k+1}. This value is added to the interpolated ORF value $J_{opt_{k+1}}\left(x_1\left(j_1^{'}\right), x_2, t_{k+1}\right)$ stored at P_{k+1}.
5. The optimal path that minimizes $L\Delta t + J_{opt_{k+1}}\left(x_1\left(j_1^{'}\right), x_2, t_{k+1}\right)$ is selected from all n_1+1 feasible paths. The optimal path destination index $j_1^{'}$ which is a grid point number of $x_1(t)$ at the $k+1$th stage is determined.
6. The procedures from (2) to (5) are executed for all grid points of the kth stage. The ORF value is stored for each stage recursively. The optimization calculation terminates if the ORF value at the initial point is obtained.

After the ORF value at the initial point has been derived, another process called approximation of optimal path selection information is applied to connect the feasible optimal points. The detailed process is described in the references [20, 21]. In these references, the evaluation results for a simple LQR design problem have indicated that PLA-DP method has a potential enough to provide an accurate solution, by comparing the two ORF values from the PLA-DP solution and the exact solution.

3.2 Gradient-Based Method

Direct optimization methods find a stationary point $\mathbf{Z}^*$ by changing the optimization variable vector $\mathbf{Z}$ such that the objective function $J(\mathbf{Z})$ is minimized. Consequently, the optimization process for unconstrained problems constructs the sequence below [10].

$$J(\mathbf{Z}_1) > J(\mathbf{Z}_2) > \cdots > J\left(\mathbf{Z}^*\right) \tag{18}$$

The gradient-based method uses a numerical gradient of both objective function and constraint vector to determine a search direction.

At first, a numerical integration scheme is introduced. The gradient-based method often adopts a multiple shooting method to avoid the serious dependency of the final states on the initial states and controls which frequently occurs in the single shooting method. The simulation of the system dynamics is executed from the intermediate states called multiple shooting nodes in each integration segment as depicted in Fig. 2. This multiple integration scheme enables to stabilize the optimal control problem even if it involves a complicated dynamic model. The definition of the discretization parameters is listed below.

$n_{\mathbf{x}_s}$: Number of multiple shooting segments

$n_{\Delta t,seg}$: Number of time fractions in one shooting segment, which corresponds to $n_{\Delta t}/n_{\mathbf{x}_s}$The system Eq. (11) is integrated using a fourth-order Runge-Kutta scheme

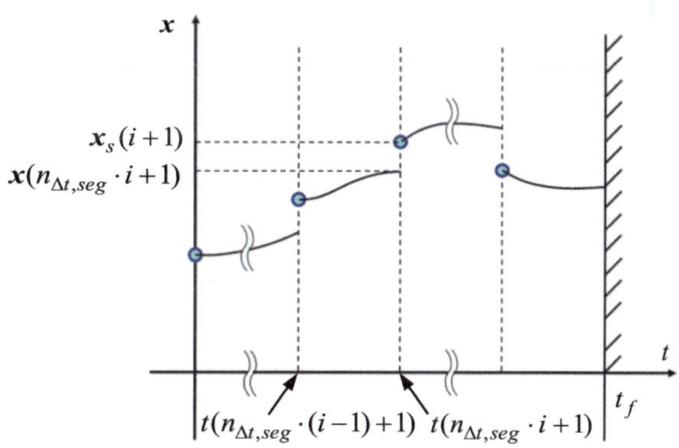

Fig. 2 Multiple shooting method

within each segment. The $(k+1)$th state vector is given by Eq. (19).

$$\boldsymbol{x}_{k+1} = \boldsymbol{x}_k + (\boldsymbol{k}_1 + 2\boldsymbol{k}_2 + 2\boldsymbol{k}_3 + \boldsymbol{k}_4)/6 \tag{19}$$

$$\boldsymbol{k}_1 = \boldsymbol{f}(x_k, u_{k+1}, t_k) \cdot \Delta t \tag{20}$$

$$\boldsymbol{k}_2 = \boldsymbol{f}(\boldsymbol{x}_k + \boldsymbol{k}_1/2, \boldsymbol{u}_{k+1}, t_k + \Delta t/2) \cdot \Delta t \tag{21}$$

$$\boldsymbol{k}_3 = \boldsymbol{f}(\boldsymbol{x}_k + \boldsymbol{k}_2/2, \boldsymbol{u}_{k+1}, t_k + \Delta t/2) \cdot \Delta t \tag{22}$$

$$\boldsymbol{k}_4 = \boldsymbol{f}(\boldsymbol{x}_k + \boldsymbol{k}_3, \boldsymbol{u}_{k+1}, t_k + \Delta t) \cdot \Delta t \tag{23}$$

As long as the multiple shooting method is used, the final states of the former segment don't coincide with the first states of the latter segment as illustrated in Fig. 2. This discontinuities, called defects, are included in the constraint vector as defect constraints $C_{\boldsymbol{x}_s}$. The $(i+1)$th defect constraint

$$C_{\boldsymbol{x}_s}(i+1) = \boldsymbol{x}\left(n_{\Delta t,seg} \cdot i + 1\right) - \boldsymbol{x}_s(i+1), \quad i = 1, 2, \cdots, n_{\boldsymbol{x}_s} \tag{24}$$

should be zero if the optimal solution could be obtained.

To obtain the optimized control input vector and state variable vector at the multiple shooting nodes under the defect and path constraints, the optimization variable vector$\boldsymbol{Z}$ and constraint vector $\boldsymbol{F}$ are defined, respectively.

$$\boldsymbol{Z}_i = \left[t_f, \boldsymbol{x}_s(i), \ \boldsymbol{u}\left(n_{\Delta t,seg}(i-1)+1\right), \cdots, \boldsymbol{u}\left(n_{\Delta t,seg} \cdot i\right), \cdots, \boldsymbol{u}(n_{\Delta t}+1)\right]^{\mathrm{T}} \tag{25}$$

$$\begin{aligned} \boldsymbol{F}_i = [&J, \ \boldsymbol{x}\left(n_{\Delta t,seg}(i-1)+1\right), \boldsymbol{y}\left(n_{\Delta t,seg}(i-1)+1\right), \cdots, \\ &\boldsymbol{x}\left(n_{\Delta t,seg} \cdot i\right), \boldsymbol{y}\left(n_{\Delta t,seg} \cdot i\right), \boldsymbol{C}_{\boldsymbol{x}_s}(i+1), \cdots, \boldsymbol{x}(n_{\Delta t}+1), \boldsymbol{y}(n_{\Delta t}+1)]^{\mathrm{T}} \end{aligned} \tag{26}$$

Note that $i = 1, 2, \cdots, n_{\boldsymbol{x}_s}$.

The $\boldsymbol{F}$ vector consists of the objective function value, state variable vector, output variable vector, and defect constraint vector. This $\boldsymbol{F}$ vector is bounded by the lower boundary $\boldsymbol{F}_L$ and upper boundary $\boldsymbol{F}_U$ to satisfy the inequality constraints on states and outputs.

$$\boldsymbol{F}_L \leq \boldsymbol{F} \leq \boldsymbol{F}_U$$

Zero vectors are assigned to eliminate the defects and to satisfy the final boundary condition. The size of $\boldsymbol{Z}$ and $\boldsymbol{F}$ is expressed by the discretization parameters.

$$n_Z = n_x n_{x_s} + n_{\boldsymbol{u}}\left(n_{\Delta t} + 1\right) + 1 \tag{27}$$

$$n_F = n_x n_{x_s} + \left(n_x + n_y\right)\left(n_{\Delta t} + 1\right) + 1 \tag{28}$$

First-Order Necessary Condition

The objective function gradient with regard to the optimization variable $\boldsymbol{Z}$ should be zero except for the first element.

$$\frac{\partial J}{\partial \boldsymbol{Z}} = \left[\frac{\partial J}{\partial t_f}, 0, 0, \cdots, 0\right] \tag{29}$$

The constraint vector gradient $\boldsymbol{G}$ is described by the Jacobian as stated in Eq. (30). In the optimization process, when a variable at a discretized time point is perturbed, only the nearby constraints are affected; therefore, the derivatives of many of the constraints with respect to the variables are zero, i.e., $\boldsymbol{G}$ normally forms a large and mostly sparse matrix.

$$\boldsymbol{G} = \frac{\partial \boldsymbol{F}}{\partial \boldsymbol{Z}} = \begin{bmatrix} \frac{\partial J}{\partial t_f} & 0 & \cdots & 0 \\ \frac{\partial F(2)}{\partial t_f} & \frac{\partial F(2)}{\partial Z(2)} & \cdots & \frac{\partial F(2)}{\partial Z(n_Z)} \\ \vdots & \vdots & \ddots & \vdots \\ \frac{\partial F(n_F)}{\partial t_f} & \frac{\partial F(n_F)}{\partial Z(2)} & \cdots & \frac{\partial F(n_F)}{\partial Z(n_Z)} \end{bmatrix} \tag{30}$$

4 Optimal Trajectory Design for Passenger Aircraft

The numerical characteristics of each method are evaluated through designing a passenger aircraft's fuel minimal optimal trajectory as an application example. Due to the limitation of computational time and memory in DP, a longitudinal optimal trajectory is designed by focusing only on aerodynamics and propulsion performance. The wind effect can be considered easily in both methods; anyhow, it's not included in the kinematic model in this paper.

4.1 Objective Function

The objective function is defined as fuel consumption from climb to descent, which is given as the time integral of fuel flow. Although flight time is another important concern, only fuel consumption is considered in the example. If the time is selected as the independent variable, where the time-based optimal control problem is formed, the objective function is defined with the terminal-free condition.

$$J = \int_{t_0}^{t_f} \mu(t)dt + \phi\left(\boldsymbol{x}\left(t_f\right), t_f\right) \tag{31}$$

4.2 *Equations of Kinematics and Dynamics*

The motion of an aircraft is generally described by the six degrees of freedom (6DOF) equations of motion, although in trajectory optimization problems, an aircraft is often approximated as a point mass because the time scale of rotational motion is sufficiently shorter than that of translational motion. Therefore, the motion of an aircraft may be described with the three degrees of freedom (3DOF) equations using the point mass approximation.

$$\frac{d\theta}{dt} = \frac{1}{(R_0 + H)\cos\phi} V \cos\gamma \sin\psi \tag{32}$$

$$\frac{d\phi}{dt} = \frac{1}{R_0 + H} V \cos\gamma \cos\psi \tag{33}$$

$$m\frac{dV}{dt} = T - D - mg\sin\gamma \tag{34}$$

$$mV\frac{d\gamma}{dt} = L - mg\cos\gamma \tag{35}$$

$$\frac{dH}{dt} = V\sin\gamma \tag{36}$$

$$\frac{dm_{fuel}}{dt} = \mu \tag{37}$$

A basic aerodynamic model is used here with a quadratic drag polar.

$$L = \frac{1}{2}\rho V^2 S C_L \tag{38}$$

$$D = \frac{1}{2}\rho V^2 S C_D \tag{39}$$

$$C_D = C_{D0} + kC_L^2 \tag{40}$$

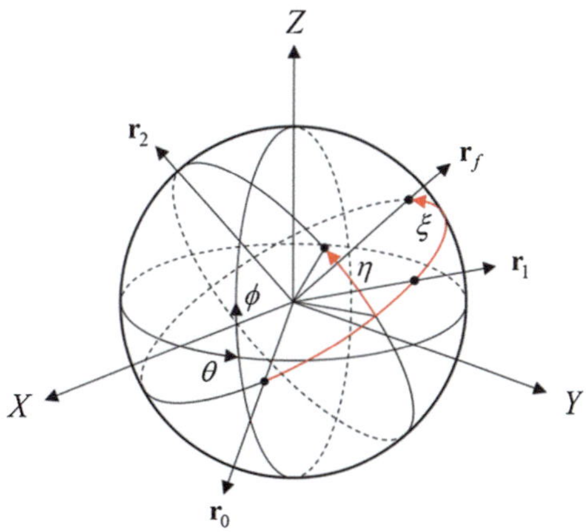

Fig. 3 Definition of downrange angle and cross range angle

The required quantities such as air density, temperature, and sound velocity are obtained by the International Standard Atmosphere (ISA) model.

As illustrated in Fig. 3, the polar coordinates are transformed from latitude and longitude (ϕ, θ) to downrange and cross range angle (ξ, η) which correspond to a great-circle course and lateral deviation. This coordinate transformation changes Eqs. (32) and (33) into the following equation:

$$(R_0 + H)\frac{d\xi}{dt} = \frac{dX}{dt} = V\cos\gamma \tag{41}$$

4.3 Trajectory Optimization

PLA-DP Method Application

The number of state variables can be reduced by setting the downrange X as the independent variable. The objective function is redefined with the terminal-fixed condition by substituting fraction of time which is obtained from Eq. (41) into Eq. (31).

$$J = \int_{X_0}^{X_f} \frac{\mu}{V\cos\gamma} dX \tag{42}$$

The state variable and control input vectors are:

$$\boldsymbol{x} = [V, \gamma, H]^{\mathrm{T}}$$

$$\boldsymbol{u} = [T, C_L]^{\mathrm{T}}.$$

The OCP in PLA-DP method is defined as minimizing the objective function (31), which leads to the following optimality condition:

$$\begin{aligned} & J_{opt_k}\left(V\left(j_{1_k}\right),\gamma\left(j_{2_k}\right),H\left(j_{3_k}\right),X(k)\right) \\ & = \min_{C_L,T}\left[\int_{X_k}^{X_{k+1}}\frac{\mu}{V\cos\gamma}dX + J_{opt_{k+1}}\left(V\left(j_{1_{k+1}}\right),\gamma\left(j_{2_{k+1}}\right),H,X\left(k+1\right)\right)\right] \end{aligned} \tag{43}$$

The following conditions should be considered:

Independent variable interval

- Downrange: $X_0 \leq X \leq X_f$

Equality constraints

- State equation: $d\boldsymbol{x}/dX = \boldsymbol{g}\left(\boldsymbol{x},\boldsymbol{u}\right)$

Inequality constraints on state variables

- Velocity: $V_{\min} \leq V \leq V_{\max}$
- Mach number: $M(V) \leq M_{MO}$
- Path climb angle: $\gamma_{\min} \leq \gamma \leq \gamma_{\max}$
- Altitude: $H_{\min} \leq H \leq H_{\max}$

Inequality constraints on control input

- Thrust: $T \leq T_{\max}$

Boundary conditions on state variables

- Initial conditions: $V(t_0) = V_0, \gamma(t_0) = free, H(t_0) = H_0$
- Final conditions: $V(t_f) = V_f, \gamma(t_f) = \gamma_f, H(t_f) = H_f$

In the transition calculation between the quantized grid points, which corresponds to the integration of system equation, the state variable value is represented by the average of two grid point values, and the small change of the state variable is approximated by finite differentiation.

$$\begin{gathered} \overline{V} = \left[V\left(j_{1_k}\right) + V\left(j_{1_{k+1}}\right)\right]/2, \quad \overline{\gamma} = \left[\gamma\left(j_{2_k}\right) + \gamma\left(j_{2_{k+1}}\right)\right]/2 \quad \overline{H} = \left[H\left(j_{3_k}\right) + H\right]/2 \\ \Delta V = V\left(j_{1_{k+1}}\right) - V\left(j_{1_k}\right), \quad \Delta\gamma = \gamma\left(j_{2_{k+1}}\right) - \gamma\left(j_{2_k}\right), \quad \Delta H = H - H\left(j_{3_k}\right) \end{gathered}$$

The control input value is approximated to be piecewise constant by a zero-order hold.

$$T^{\circ} = T(X), X \in \left[X_k, X_{k+1}\right], \quad C_L{}^{\circ} = C_L(X), X \in \left[X_k, X_{k+1}\right]$$

These assumptions lead to the control inputs in the transition, which are used to calculate the ORF value.

$$T^{\circ} = D\left(\overline{V}, \overline{H}\right) + mg\sin\overline{\gamma} + m\overline{V}\cos\overline{\gamma}\frac{\Delta V}{\Delta X} \tag{44}$$

$${C_L}^{\circ} = \frac{L\left(\overline{V}, \overline{\gamma}\right)}{\rho\left(\overline{H}\right)\overline{V}^2 S/2} \tag{45}$$

Although velocity and flight-path climb angle are controlled directly by the two control inputs, altitude cannot be controlled directly. Thus, the feasible solution is distributed on the $V-\gamma$ surface, and the feasible altitude is consequently not assigned on the quantized grid points. This typical "menace of the expanding grid" can be resolved by linear interpolation denoted in Eq. (17), which is a core concept of PLA-DP method. The feasible altitude points are derived analytically by one of the state Eq. (46) which is given by Eqs. (36) and (41).

$$H = H\left(j_{3_k}\right) + \Delta X\tan\overline{\gamma} \tag{46}$$

The state variables and independent variable are quantized at even interval to set an equidistant calculation grid.

Gradient-Based Method Application

The same optimal control problem is solved by using the same independent variable, downrange. The downrange is divided equally into$n_{\Delta X}+1$ points.

$$X_k = X_0 + \left(X_f - X_0\right)(k-1)/n_{\Delta X}, \quad k = 1, 2, \cdots, n_{\Delta X}+1 \tag{47}$$

$n_{\Delta X}$: Number of downrange steps.

The optimization vector and constraint vector are written down with state and control inputs.

$$\begin{aligned} \boldsymbol{Z} = \big[& X\left(n_{\Delta X}+1\right), V_s(1), \gamma_s(1), H_s(1), \delta T(1), C_L(1), \cdots, \\ & \delta T\left(n_{\Delta X,seg}\right), C_L\left(n_{\Delta X,seg}\right), \cdots, \delta T\left(n_{\Delta X}+1\right), C_L\left(n_{\Delta X}+1\right)\big]^{\mathrm{T}} \end{aligned} \tag{48}$$

$$\begin{aligned} \boldsymbol{F} = \big[& m_{fuel}\left(n_{\Delta X}+1\right), V(1), \gamma(1), H(1), n_z(1), \cdots, V\left(n_{\Delta X,seg}\right), \gamma\left(n_{\Delta X,seg}\right), \\ & H\left(n_{\Delta X,seg}\right), n_z\left(n_{\Delta X,seg}\right), \\ & V\left(n_{\Delta X,seg}+1\right) - V_s(2), \gamma\left(n_{\Delta X,seg}+1\right) - \gamma_s(2), H\left(n_{\Delta X,seg}+1\right) - H_s(2), \\ & \cdots, V\left(n_{\Delta X}+1\right), \gamma\left(n_{\Delta X}+1\right), H\left(n_{\Delta X}+1\right), n_z\left(n_{\Delta X}+1\right)\big]^{\mathrm{T}} \end{aligned} \tag{49}$$

Two vectors $\boldsymbol{Z}$ and $\boldsymbol{F}$ are used to calculate the numerical gradient $\boldsymbol{G}$ by finite differentiation in the trajectory optimization tool. This gradient is provided to the NLP optimizer Interior Point OPTimizer (IPOPT) [22] to compute the search direction so as to satisfy the first-order necessary condition denoted by Eq. (29). The optimization tool used here has been developed in MATLAB and is based on a student course given at the Institute of Flight System Dynamics, Technische Universität München. It uses a MATLAB-executable (mex) file using MATLAB Coder for fast computation.

4.4 Aircraft Performance Model

The performance of aircraft is calculated by the BADA (Base of Aircraft Data) model version 3.11, which is developed and maintained by the European Organization for the Safety of Air Navigation (EUROCONTROL) [23]. BADA includes parametric models for multiple aircraft, such as an aerodynamic model C_{D0} and k, maximum thrust model, and fuel flow model, which are used for performance calculation. The operational limitations of each aircraft, such as its maximum operating Mach number (MMO) and maximum operating altitude (HMO), are also provided. The fuel flow calculated by the BADA model has been evaluated with reference to an airliner's on-board flight data, indicating that the model is sufficiently accurate for trajectory optimization calculations. The colored contour in Fig. 4 shows specific range (SR), which presents the flight range per unit mass of fuel consumption, calculated with the BADA model and the standard weight of a passenger aircraft operated in Japanese domestic airspace. The maximum SR,147[m/kg], is achieved by the maximum Mach number ($M = 0.84$) and maximum operating altitude ($H = 43,100$[ft]) at the corner of the flight envelope.

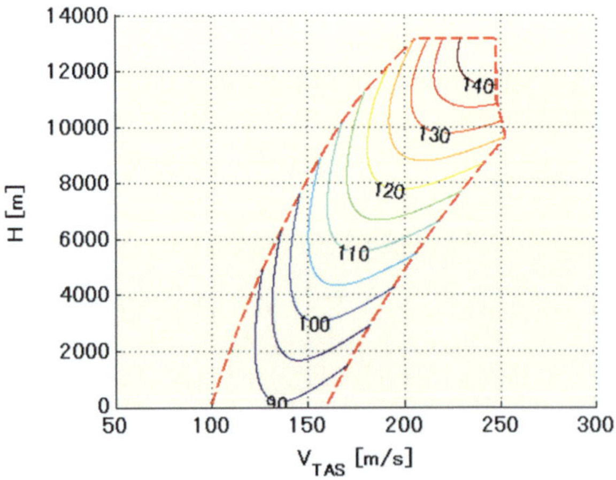

Fig. 4 Specific range (*SR* [m/kg]) in the flight envelop

4.5 Numerical Results

The fuel minimal optimal trajectory is designed by PLA-DP method and gradient-based method. The calculation conditions for the independent variable, state variables, inequality constraints, and boundary conditions are listed in Table 1. The downrange interval is set assuming the distance between Tokyo International Airport and Fukuoka International Airport in Japan. The final path climb angle of -3 [deg] is assigned to intercept a glideslope continuously.

The optimal trajectories gained by the PLA-DP method and the gradient-based method are shown in Figs. 5, 6, 7, 8, 9, 10, 11, and 12, where the circle marker represents the multiple shooting nodes used in gradient-based method. The two kinds of trajectory indicate an extremely good agreement. The difference of fuel consumption and flight time is about 1[%] as shown in Table 2.

From a perspective of flight dynamics, the characteristics of each flight phase are stated succinctly. The aircraft climbs by the maximum thrust which depends on altitude with increasing the velocity rapidly. This maximum thrust determines the optimal velocity and flight-path climb angle such that the rate of climb is maximized

Table 1 Calculation conditions

Downrange interval [m]	$X_0 = 0$	$X_f = 866.3 \times 10^3$	$\Delta X_{DP} = 28.88 \times 10^3$, $\Delta X_{Gradient} = 2.89 \times 10^3$	
Velocity [m/s]	$V_{\min} = 140$	$V_{\max} = 260$	$\Delta V = 2$	
Path climb angle [deg]	$\gamma_{\min} = -5$	$\gamma_{\max} = 5$	$\Delta\gamma = 0.2$	
Altitude [m]	$H_{\min} = 3,000$	$H_{\max} = 13,000$	$\Delta H = 100$	
Downrange [m]	$X_{\min} = X_0$	$X_{\max} = X_f$		
Mach number	$M_{MO} = 0.84$			
Initial condition	$V_0 = V_{\min}$	$\gamma_0 = free$	$H_0 = H_{\min}$	$X_0 = X_{\min}$
Final condition	$V_f = V_{\min}$	$\gamma_f = -3[\text{deg}]$	$H_f = H_{\min}$	$X_f = X_{\max}$

Fig. 5 Altitude

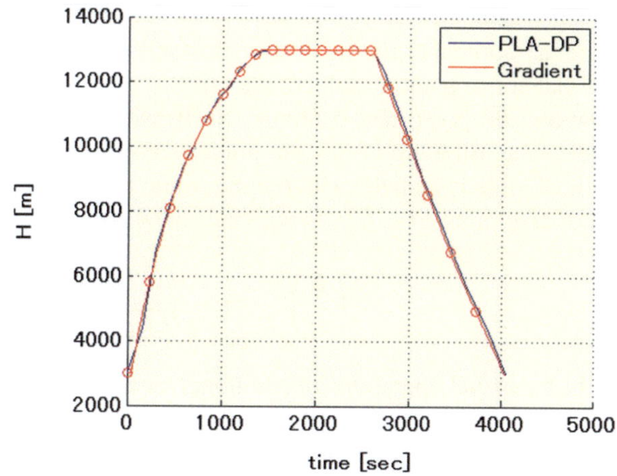

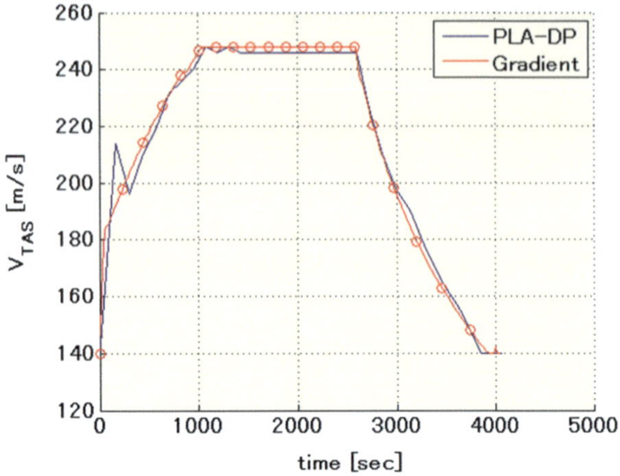

Fig. 6 Velocity

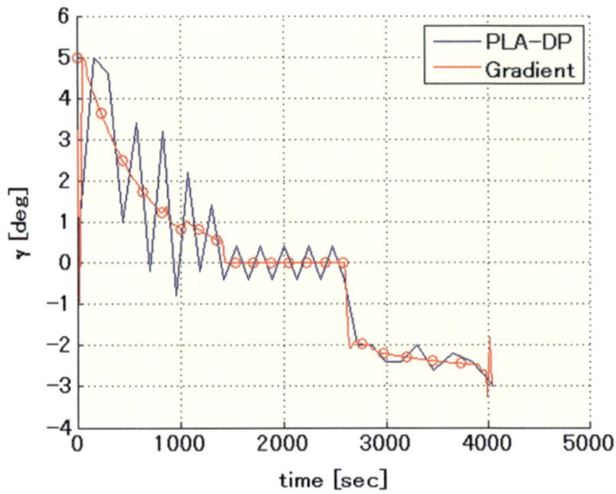

Fig. 7 Flight-path climb angle

to get an optimal cruise point in a short amount of time. Once the aircraft gets the optimal cruise point, it reduces the throttle lever position to 84.2 [%] with the fuel flow of 1.68 [kg/s] and continues to fly with maintaining the maximum specific range value till a top of descent point is reached. In the descent phase, the typical optimal glide is realized with the minimum thrust setting. The achieved maximum lift to drag ratio, 17.4, agrees with the analytical value given by Eq. (50).

$$\left(\frac{L}{D}\right)_{\max} = \frac{1}{2\sqrt{KC_{D0}}} = 17.4286 \tag{50}$$

In the actual operation, an additional term with a parameter called cost index is often included in the objective function to evaluate the flight time [1–7]. If the cost index is set so as to increase the flight time, the aircraft starts to descend earlier to ensure a longer descent phase. In the opposite case, the top of descent point moves to

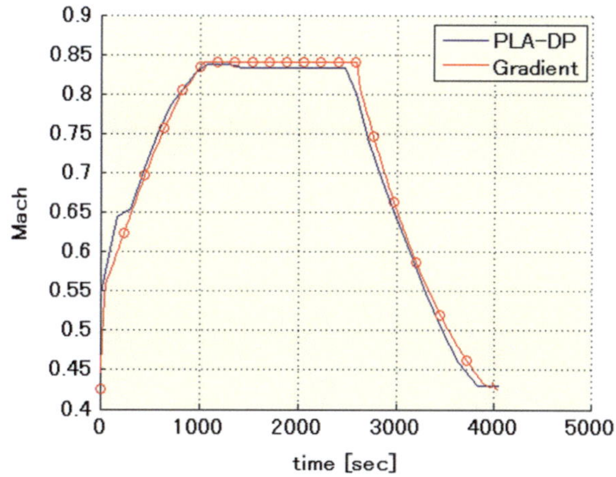

Fig. 8 Mach number

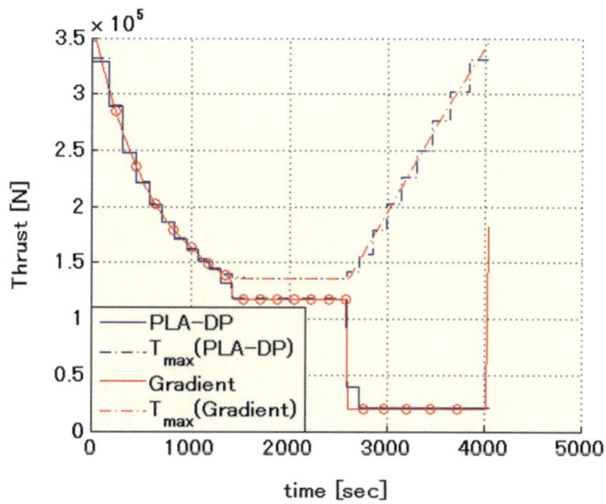

Fig. 9 Thrust

behind. The total fuel consumption surely increases for both cases, and the feasible solution is distributed on a Pareto front. This fact is confirmed by several analysis results with Dynamic Programming in the previous literature [2]. It is also possible to consider the flight time assigning a cost index in gradient-based method.

An oscillation is observed in the optimal trajectory of flight-path angle derived by PLA-DP method as can be seen in Fig. 7. This oscillation inevitably arises by calculating the objective function value using a mean value between two grid points; however, two measures can be taken to eliminate the unfavorable oscillation. One is improving the integration accuracy by dividing the transition into two or more parts, and the other is minimizing the total amount of path angle change by adding a term such as integral of squared path angle rate into the objective function.

The computational time is shown in Fig. 13 with bar graph. The specification of the used machine is listed below.

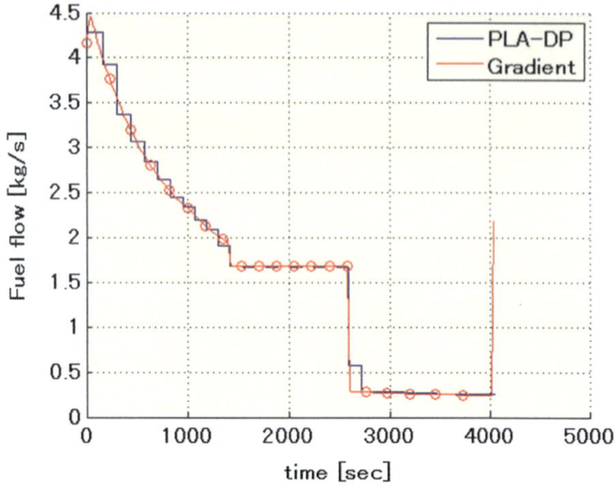

Fig. 10 Fuel flow

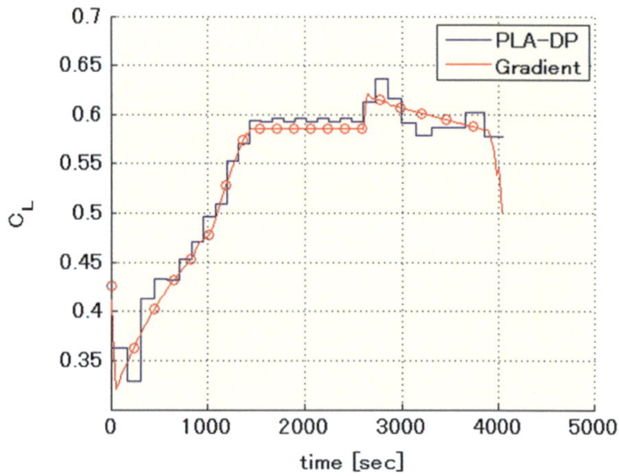

Fig. 11 Lift coefficient

Machine Specification

- OS: Windows 8.1 (64-bit)
- CPU: Intel Core i7-4510U@2.00GHz
- RAM: 8.0 GB

The computational time taken by Dynamic Programming method was 4505 seconds; on the other hand, it was 359 seconds in the gradient-based optimization. Both numbers hold for the implementation used here, of course, and may not be transferred to another problem or implementation. The computational time spent on gradient-based calculation depends heavily not only on the complexity of the problem such as the number of inequality constraints or boundary conditions but on the implementation of the code. Therefore, it is possible to get much shorter computational time; anyway, this short amount of time is worth a special remark as a reasonable one for a practical usage.

Fig. 12 Lift to drag ratio

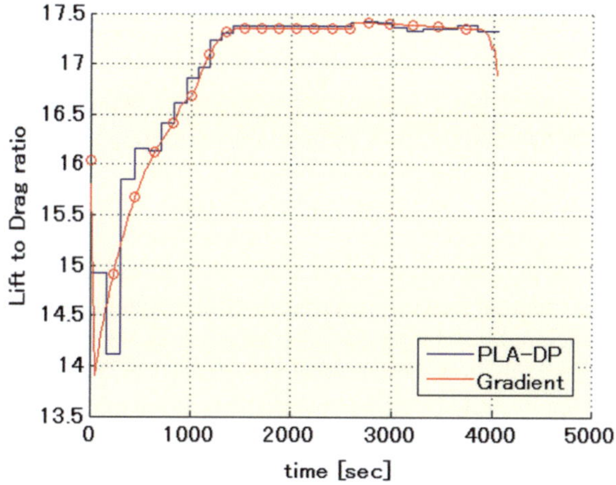

Table 2 Numerical comparison

	PLA-DP (a)	Gradient-based method (b)	Relative difference ((b)−(a))/(a)
Fuel consumption [kg]	6534.1	6447.1	−0.0133
Flight time [s]	4055.4	4039.1	−0.0040

Fig. 13 Computational time for each method

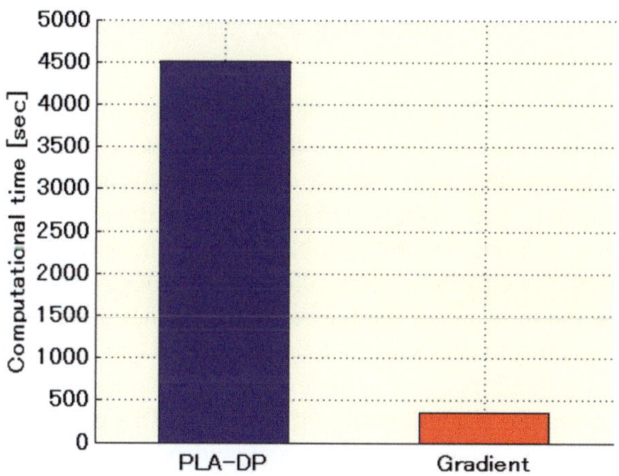

Regarding the applicability of the optimization methods including a practical model, a complicated model and tabular data can be easily applied in Dynamic Programming method. On the other hand, in gradient-based method, a model should be described with a continuous function, and a tabular data must be arranged so as to construct a smooth surface. These continuous models and smoothed data encourage gradient-based method to produce a smooth optimal trajectory; nevertheless, the modeling process should be implemented not to detract the characteristics of the original model and data.

5 Conclusion

This paper has evaluated the usability of two direct methods for solving optimal control problems, PLA-DP method and gradient-based method, on a practical trajectory optimization problem toward the realization of more efficient operations in the ATM system. The computational features for each method have been investigated in terms of convergence to a realistic solution, computational time, and applicability for the practical model. As a practical application within the scope of TBO concept, fuel minimal longitudinal trajectory has been designed with a realistic performance model for all flight phases. The results indicate that the optimal trajectories obtained by each method show a great agreement, i.e., an optimal solution close to a global optimum could be gained by gradient-based method in this case. The other findings are collected below.

Advantages in PLA-DP Method

1. The computational time is predictable since the total amount of calculation is decisive.
2. A complicated model can be easily applied.

Advantages in Gradient-Based Method

1. A fast computation is possible even if a precise integration scheme is used.
2. A smooth optimal trajectory can be obtained by the numerical integration scheme.

Dynamic Programming has been successfully used as one trajectory optimization analysis method to evaluate the potential benefits of CARATS, where the fuel consumption and flight time are compared between two kinds of values on the optimal trajectory and on the real trajectory reconstructed from secondary surveillance radar data or airliner's flight data. Nevertheless, the aircraft dynamics could not be considered explicitly because the allowable number of optimization variable was restricted by the three major hindrances. On the other hand, the gradient-based method has capability to deal with the longitudinal and lateral/directional motion, and the results show that it is possible to design a longitudinal optimal trajectory using a realistic performance model within 359 seconds. This great advantage of the method might encourage generating an optimal guidance law within a reasonable time on board, whereby the aircraft will be able to fly on an optimal trajectory more precisely.

The gradient-based method is expected to be utilized as a promising trajectory optimization method for designing a practically preferable reference trajectory toward the realization of more efficient operation in the ATM system.

Acknowledgments This research was conducted while the author studied as a visiting researcher at the Institute of Flight System Dynamics, Technische Universität München. The four researchers who belong to the flight optimization group, Dipl.-Ing. Matthias Bittner, Dipl.-Ing. Matthias Rieck, M. Sc. Benedikt Grüter and M. Sc. Johannes Diepolder, and the Institute Chair, Professor Dr.-Ing. Florian Holzapfel, gave me worthful comments and advice. I would like to express my cordial appreciation to their support.

References

1. Miyazawa, Y., Matsuda, H., Shigetomi, S., Harada, A., Kozuka, T., Mickramasinghe, N. K., Brown, M. and Fukuda, Y.: Potential Benefits of Arrival Time Assignment, 11th USA/EUROPE Air Traffic Management R&D Seminar, Lisbon, 2015.
2. Wickramasinghe, N. K., Brown, M., Fukuda, Y., Harada, A., and Miyazawa, Y.: Correlation between Flight Time and Fuel Consumption in Airliner Flight Plan with Trajectory Optimization, AIAA Guidance, Navigation, and Control Conference, AIAA-2015-1318, Florida, 2015.
3. Harada, A., Kozuka, T., Miyazawa, Y., Wickramasinghe, N. K., Brown, M. and Fukuda, Y.: Analysis of Air Traffic Efficiency using Dynamic Programming Trajectory Optimization, 29th International Congress of the Aeronautical Sciences, St. Petersburg, 2014.
4. Miyazawa, Y., Wickramasinghe, N. K., Harada, A. and Miyamoto, Y.: Dynamic Programming Application to Airliner Four Dimensional Optimal Flight Trajectory, AIAA Guidance, Navigation and Control Conference, AIAA-2013-4969, Boston, 2013.
5. Wickramasinghe, N. K., Miyamoto, Y., Harada, A., Kozuka, T., Shigetomi, S., Miyazawa, Y., Brown, M. and Fukuda, Y.: Flight Trajectory Optimization for Operational Performance Analysis of Jet Passenger Aircraft, *Trans. JSASS Aerospace Technology Japan*, **12** (2014), pp. a17-a25.
6. Wickramasinghe, N. K., Harada, A., Totoki, T., Miyamoto, Y. and Miyazawa, Y.: Flight Trajectory Optimization for Modern Jet Passenger Aircraft with Dynamic Programming, *Air Traffic Management and Systems*, Lecture Notes in Electrical Engineering, **290** (2014), pp. 87-104.
7. Miyamoto, Y., Wickramasinghe, N. K., Harada, A., Miyazawa, Y. and Funabiki, K.: Analysis of Fuel-Efficient Airliner Flight via Dynamic Programming Trajectory Optimization, *Trans. JSASS Aerospace Technology Japan*, **11** (2013), pp. 93-98.
8. Bittner, M., Fleischmann, B., Richter, M. and Holzapfel, F.: Optimization of ATM Scenarios Considering Overall and Single Costs, 6th International Conference on Research in Air Transportation Istanbul, 2014.
9. Harada, A., Miyazawa, Y. and Funabiki, K.: Accuracy evaluation of an aircraft performance model with airliner flight data, *Trans. JSASS Aerospace Technology Japan*, **11** (2013), pp. 79-85.
10. Betts, J.T.: *Practical Methods for Optimal Control Using Nonlinear Programming*, Society for Industrial and Applied Mathematics, 2001, pp. 81-178.
11. Bellman, R., Dynamic Programming, Princeton Univ. Press, 1957.
12. Bryson, A. E. and Ho, Y. C.: Applied Optimal Control, Ginn and Company, 1969, pp. 128-147.
13. Schulz, W. and Schulze H.-K., "A Method of Dynamic Programming and its Application to Optimization Problems of Flight Mechanics," NASA Technical Translation, NASA TT F-483, 1967.
14. Jaroslaw, S.S., "Overcoming the Bellman's Curse of Dimensionality in Large Optimization Problems," NASA Technical Memorandum 102662, 1990.
15. Luss, R., Iterative Dynamic Programming, Chapman & Hall/CRC, London, UK, 2000.
16. Miyazawa, Y., Harada, A., Wickramasinghe, N. K. and Miyamoto, Y.: Effect of Jet Passenger Aircraft Performance Model on the Optimal Periodic Cruise Maneuver, in Japanese, *JSASS Kouku-Uchu Gijutsu*, **12** (2013), pp. 99-105.
17. Miyazawa, Y., Harada, A., Kawaguchi J, Ninomiya T., Suzuki, H. and Tomita, H., "Dynamic Programming Trajectory Optimization and its Application to D-SEND#2 Low Sonic-boom Research Project," AIAA 2012-4827, AIAA Guidance, Navigation and Control Conference, Minneapolis, 2012.
18. Harada, A. and Miyazawa, Y., "Dynamic Programming Applications to Flight Trajectory Optimization," 19th IFAC Symposium on Automatic Control in Aerospace, Würzburg, 2013.
19. Lapidus, L. and Luus, R.: Optimal Control of Engineering Processes, Braisdell, Waltham, Massachusetts, 1967, pp. 84-86.

20. Matsuda, H., Harada, A. and Miyazawa, Y.: Trajectory Optimization using Piecewise Linear Approximation Dynamic Programming, in Japanese, *JSASS Kouku-Uchu Gijutsu*, **14** (2015), pp. 33-41.
21. Harada, A., Matsuda, H. and Miyazawa, Y.: Dynamic Programming Trajectory Optimization by Piecewise Linear Approximation, AIAA Guidance, Navigation, and Control Conference, AIAA-2015-1075, Florida, 2015.
22. Wächter, A. and Biegler, L. T.: On the implementation of an interior-point filter line-search algorithm for large-scale nonlinear programming, Mathematical Programming, **106**, 1 (2006), pp. 25-57.
23. Eurocontrol Experimental Center: *User Manual for the Base of Aircraft Data (BADA) Revision 3.11*, EEC-Technical-Report-130416, April, 2013.

Optimization-Based Performance Assessment on 4D Trajectory-Based Operations with Track Data

Navinda Kithmal Wickramasinghe, Mark Brown, Sachiko Fukushima, and Yutaka Fukuda

Abstract The enhancement of air traffic management (ATM) system performance by management of aircraft trajectories, commonly referred to as 4D trajectory-based operations (TBO), is one of the key technologies in the Japan Civil Aviation Bureau's "Collaborative Actions for Renovation of Air Traffic Systems (CARATS)" plan. The release of "CARATS Open Data," aircraft track data for all scheduled commercial instrument rule flights in Japan's en route airspace, has enabled the understanding of performance in present flight operations on a broad scale and assesses the potency of possible benefits toward a futuristic ATM system. This paper focuses on potential benefits in a 4D TBO system through operational performance on existing flight operations based on trajectory optimization. A trajectory optimization model developed by the authors is used to investigate the potential benefits by exerting the maximum performance of the aircraft in respect to highly regulated restrictions used in current flight procedures. Quantitative results show that weather conditions have a significant impact on conventional operational performance. Optimized results denote that substantial benefits could be obtained by more relaxed flight planning which vary according to arrival time assignment and weather conditions.

Keywords Open data • Trajectory optimization • Operational performance

---

N.K. Wickramasinghe (✉) • M. Brown • S. Fukushima
Air Traffic Management Department, Electronic Navigation Research Institute (ENRI), Tokyo, Japan
e-mail: navinda@mpat.go.jp

Y. Fukuda
Navigation Systems Department, Electronic Navigation Research Institute (ENRI), Tokyo, Japan

Electronic Navigation Research Institute, *Air Traffic Management and Systems II*,
Lecture Notes in Electrical Engineering 420, DOI 10.1007/978-4-431-56423-2_6

1 Introduction

The enhancement of air traffic management (ATM) system performance by management of aircraft trajectories, commonly referred to as 4D trajectory-based operations (TBO) concept, is one of the key technologies in the Japan Civil Aviation Bureau (JCAB)'s "Collaborative Actions for Renovation of Air Traffic Systems (CARATS)" research plan [1]. To enable wider participation of research bodies, industries, and universities in research to support the CARATS project, JCAB has initiated a program to release the so-called CARATS Open Data (henceforth referred to as open data) since February 2015, a series of aircraft track data for all scheduled commercial instrument flight rule flights in Japan's en route airspace derived from air route surveillance radars (ARSR) [2]. ENRI's role in this program is to provide data processing services to the JCAB and act as a collaborator between JCAB and data users in promoting research and development projects based on the open data. In parallel, with the objective of contributing toward the CARATS program, ENRI has begun a 4-year project aimed at clarifying the concept of TBO. The study aims to derive a future "full 4D TBO" operation environment for Japanese airspace by evaluating simulation-based benefit clarification, potential issue identification, and solution recommendations [3], targeting the anticipated national air traffic volume in year 2030. Development of a trajectory optimizer with the collaboration of Kyushu University is one of the main objectives of this study with a projected stage to extend the scope by upgrading the optimizer to generate optimal conflict-free trajectories for a future ATM system. The distribution of open data provides an excellent platform for the authors' research project to investigate various aspects of benefits in 4D TBO operations while reviewing the degree of performance in conventional ATC operations.

The significance of quantitative evaluations based on big data analyses is to understand the performance of present flight operations on a broad scale and assess the potency of possible benefits toward a future ATM system. Bourgois et al. provide a statistical perspective on the significance of big data distribution toward enhancing R&D projects in the aviation industry [4]. Dobruszkes also emphasizes the importance of data availability to create awareness on transport geography [5]. Research studies focused on major airports to understand the performance trend and find solutions to improve its operational efficiency are abundant in the United States and Europe [6–9]. On the contrary, benefit assessments concentrating on specific airports and/or specific flight phases are conducted by authors and in related studies [10–12], but a quantitative evaluation on potential benefits considering the entire system is a new approach to review the performance assessment in a comprehensive scale. The necessity to explicitly express the potential benefits of a 4D TBO system in targeting the objectives of CARATS and promotion of the CARATS Open Data program has motivated the authors to implement this research.

This paper focuses on potential benefits of a 4D TBO system from the viewpoint of trajectory optimization based on the performance of conventional flight operations. Flight parameters are estimated for several major aircraft types utilized in domestic operations, by applying meteorological data provided by the Japan Meteorological Agency (JMA) and aircraft performance data from the Base of Aircraft Data (BADA) aircraft performance model of Eurocontrol to time histories of aircraft position acquired from open data. The noise included in the open data due to errors from tracking system, data resolution, data integration, etc. is reduced by a smoothing technique, and the validation of the process is reviewed through an analytical data comparison. A trajectory optimization model proposed in authors' previous studies [13, 14] is used to generate fuel-optimal trajectories to address the performance assessment of conventional procedures and emphasize suggestions to improve critical areas in the current system in terms of efficiency. Optimized results are investigated with/without arrival time assignment to understand the variations in achieved benefits due to the trade-off between fuel consumption and flight time/flight range and meteorological conditions.

The structure of this paper consists of an introduction on the characteristics of the open data in section "CARATS open data", an introduction on the utilized models in this research in section "Utilized models", explanation of scenario on problem setting in section "Research scenario", analysis on operational performance analysis in section "Operational performance", and potential benefit assessment in section "Potential benefit assessment" and concludes with conclusions in section "Conclusion".

2 CARATS Open Data

The distributed data includes time histories of 3D position data of about 150,000 flights with virtual call signs assigned at the initial stage of data processing. Table 1 shows the properties of the open data. A typical scenario of daily flight operations in Japan is illustrated in Fig. 1, in which the color distinction resembles the flight altitude. Areas margined by blue-dotted lines resemble the four area control center (ACC) coverage areas from north to south, composed of Sapporo ACC, Tokyo ACC, Fukuoka ACC, and Naha ACC. The Kanto area (mainly around Haneda and Tokyo international airports) is the most highly congested airspace in Japan. Generally, the distance and heading for each aircraft are tracked by ARSR at a 10-s interval within

Table 1 Properties of open data

Parameter	Dimensions/characteristics
Flight time	hh:mm:ss/Japan standard time (JST)
Flight number	Virtual number allocated during data processing
Latitude	deg
Longitude	deg
Altitude	ft/pressure altitude
Type	ICAO designator code

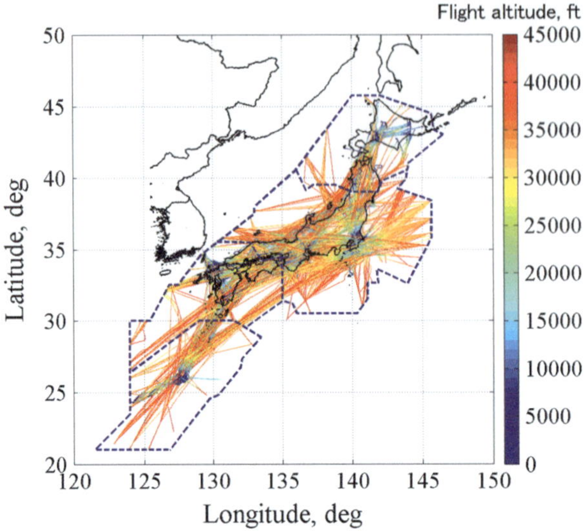

Fig. 1 Daily flight operations in Japanese airspace

each of its coverage area. Then, a radar data processing (RDP) system converts the raw data to latitude and longitude values by using a coordinate transformation process and displays aircrafts' movement on the air traffic control panels enabling the air traffic controllers to man the air traffic. These data are further processed for the open data distribution at ENRI. A detailed review on the data distribution, characteristics, and its promotion in the field of ATM research can be found in Ref. [15].

3 Utilized Models

The following models are applied to review the operational performance and benefit estimation with respect to above-described open data. Information regarding airways, waypoints, aerodromes, etc. is provided by the Japan Aeronautical Information Service Center (AIS Japan) [16].

3.1 Meteorological Model

Meteorological data is acquired from the global spectral model (GSM) numerical weather prediction (NWP) data for the Japan region distributed by the JMA [17]. A linear interpolation along the time axis and a weighted interpolation along the aircraft's 3D position are used on physical elements to extract weather data at the time stamp of each flight data point in the analysis of operational performance. In case of trajectory optimization, weather data of which the initial prediction time is closest to the departure time of the corresponding flight is used with the assumption that weather conditions do not change significantly during the entire flight period. This assumption is plausible because the weather prediction data update has a 6-h time window, which is considerably large compared to even the longest domestic flight, which is just above 2 h.

3.2 *Aircraft Performance Model*

The BADA aircraft performance model data (ver. 3.11) developed and maintained by Eurocontrol is utilized to address the subjected aircrafts' performance [18]. Reference mass provided in the BADA model is applied in the study, as the open data does not include data on aircraft mass. Also, operational constraints defined in the BADA model are utilized in the trajectory optimization model to obtain plausible optimal trajectories.

3.3 *Trajectory Optimization Model*

The authors have developed a trajectory optimization model to review the potential benefits of 4D TBO such as fuel consumption and flight time by enabling the pilot to exert the maximum performance of the aircraft on a 4D trajectory platform. The detailed explanation of the model is mentioned in previous studies, while this section provides a brief introduction for the reader's feasibility.

3.3.1 Equations of Motion

Point mass equations are defined for the aircraft's 3D translational motion as shown in Eqs. (1, 2, 3, 4).

$$\frac{\mathrm{d}\theta}{\mathrm{d}t} = \frac{1}{(R_0 + H)\cos\phi}(V_{\mathrm{TAS}}\cos\gamma_a\sin\psi_a + W_x) \tag{1}$$

$$\frac{\mathrm{d}\phi}{\mathrm{d}t} = \frac{1}{R_0 + H}\left(V_{\mathrm{TAS}}\cos\gamma_a\cos\psi_a + W_y\right) \tag{2}$$

$$\frac{\mathrm{d}H}{\mathrm{d}t} = V_{\mathrm{TAS}}\sin\gamma_a \tag{3}$$

$$m\frac{\mathrm{d}V_k}{\mathrm{d}t}\cos(\gamma_a - \gamma)\cos(\psi_a - \psi) = T - D - mg\sin\gamma_a \tag{4}$$

ϕ	Latitude	g	Gravitational acceleration
θ	Longitude	m	Aircraft mass
H	Altitude	V_k	Inertial speed
V_{TAS}	True airspeed	T	Engine thrust
γ	Flight path angle	D	Aerodynamic drag
ψ	Azimuth angle	**Subscripts**	
R_0	Earth radius	a	Respect to air flow
t	Time	x	Zonal component
W	Wind	y	Meridional component

The cosine component of the inertial speed vector is added to Eq. (4) in order to consider its effect on the aircraft's motion along the direction of thrust vector, which is parallel to the true speed vector. The 3D position and speed of the aircraft are considered as the state variables. These are discretized, and the dynamic programming (DP) method is applied to generate the fuel optimal trajectories.

3.3.2 Application of DP Method

The state space grid is discretized, and the optimal trajectory is calculated by finding the optimal segment of each transition within the grid as shown in Eq. (5). The performance index to be minimized is defined in Eq. (6). The cost index (CI), which directly corresponds to the trade-off between fuel consumption and flight time [19], is used through conversion of dimensions to redefine the performance index for feasibility as given in Eqs. (7) and (8). Figure 2 illustrates the application of DP

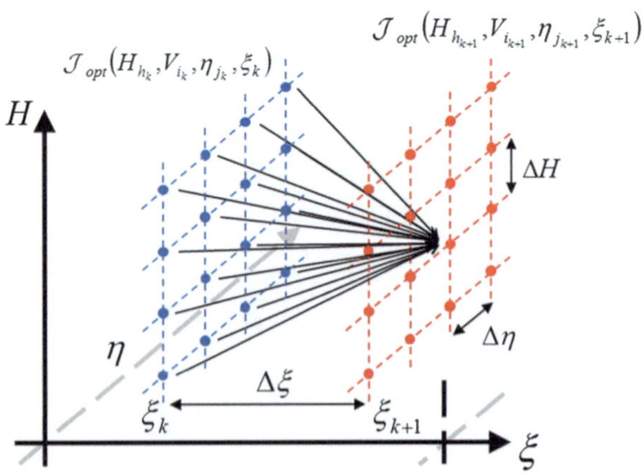

Fig. 2 Application of DP method

method:

$$J_{opt}\left(H_{h_{k+1}}, V_{i_{k+1}}, \eta_{j_{k+1}}, \xi_{k+1}\right) = \min_{\substack{h_k \to h_{k+1} \\ i_k \to i_{k+1} \\ j_k \to j_{k+1}}} \left[H_{h_k}, V_{i_k}, \eta_{j_k}, \xi_k + FF|_{\xi_k}^{\xi_{k+1}} \Delta t\right] \quad (5)$$

$$\min\ J = \int_{t_0}^{t_f} C_{\text{fuel}} \cdot FF(t)\ \mathrm{d}t + \int_{t_0}^{t_f} C_{\text{time}}\ \mathrm{d}t \quad (6)$$

$$\min\ J = \mu\left(t_f - t_0\right) + \int_{t_0}^{t_f} FF(t)\ \mathrm{d}t \quad (7)$$

$$CI = \frac{C_{\text{time}}}{C_{\text{fuel}}} = 79.37a \quad (8)$$

ξ	Along-track angle	μ	Weighting parameter
V	Calibrated airspeed	η	Cross-track angle
FF	Fuel flow	J	Performance index
C_{time}	Time cost	C_{fuel}	Fuel cost
Subscripts			
h, i, j	Arbitrary waypoints along state variable axes	k	Arbitrary waypoint along independent variable axis
0	Initial	f	Final
opt	Optimal		

The weighting parameter enables various settings of the optimizer to generate trajectories optimized only for fuel or trajectories optimized for fuel with flight time constraints. This capability is used to evaluate the characteristics of the aircraft performance in conventional procedures and investigate the impact of 4D TBO operations in a future ATM system. The next section provides a detailed explanation of the scenario of this study.

4 Research Scenario

This section reviews the analysis flow of the research study including the data preparations for the performance analysis. The analysis is based on one third of daily domestic flights operated in the Japanese airspace by three main aircraft types. Furthermore, two arbitrary days are selected, each from summer and winter seasons to review the effect of weather conditions on the operational performance. In the analysis, aircraft A is a twin-engine, narrow body, short to medium range jet aircraft; aircraft B is a twin-engine, wide body, medium to long-range jet aircraft; and aircraft

C is a twin-engine, wide body, long-range jet aircraft. The number of aircraft used in the study and its main characteristics are shown in Tables 2 and 3.

4.1 Analysis Flow

Figure 3 visualizes the analysis flow in four steps:

- Step 1: Air data estimation

  The 3D position and time of each flight are extracted, and air data such as true airspeed, calibrate airspeed, and Mach number are estimated by using JMA weather data.

Table 2 Number of aircraft used in the analysis

Aircraft type	2012/05	2013/01
A	445	486
B	289	273
C	169	153
Total	903	912

Table 3 Physical characteristics of subjected aircraft

	Aircraft type		
Physical property	A	B	C
Aircraft mass [ton]	65.3	154.5	208.7
Wing area [m^2]	124.65	283.35	427.82
Maximum operating speed [kt]	290	310	310
Maximum operating Mach number	0.78	0.80	0.84
Stall speed (for reference mass) [kt]	149	167	149

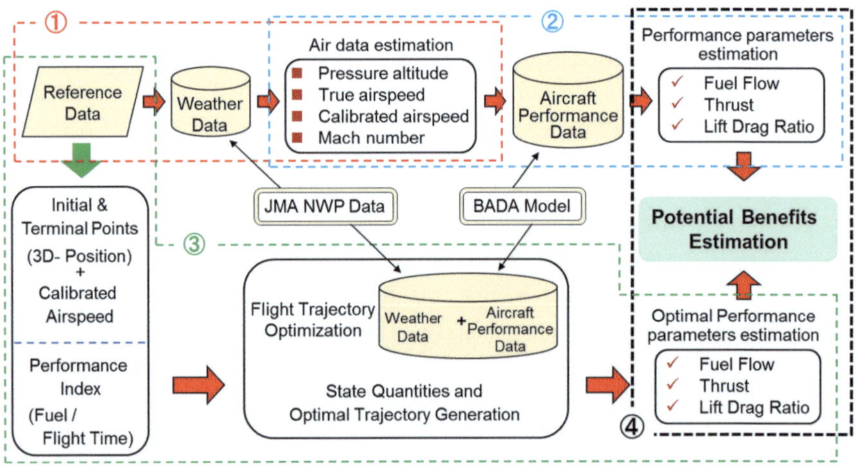

Fig. 3 Analysis flow

- Step 2: Estimation of performance parameters
  Estimated air data are utilized with BADA data to estimate each flight's performance such as fuel flow, thrust, and lift-to-drag ratio.
- Step 3: Trajectory optimization
  The required initial and terminal boundary conditions are extracted from each flight to generate fuel optimal trajectories with given CI values by using identical weather data and BADA data used in step 1.
- Step 4: Estimation of potential benefits
  Estimated performance parameters from step 2 and optimal performance parameters from step 3 are compared to investigate the impact of 4D TBO application in the performance assessment.

4.2 Data Preparation

Usually, track data includes noise due to data integration methods used according to different geographical locations of radar data sites [15]. Therefore, a smoothing technique is used to process the open data before estimating the performance parameters. A smoothing algorithm proposed by Bach et al. [20] to treat raw radar data is used here to remove odd data points and interpolate uneven data recordings followed by a zero-phase finite impulse filter. To understand the validity of this combination, a series of flight data logged in a commercial GPS receiver, mentioned in Ref. [13], is used to compare the performance parameter estimation by assuming the position data recorded in the GPS receiver are highly precise.

Figures 4a, 4b, 4c and 4d illustrates the results of an example for compared performance parameters, namely, true track angle, engine thrust, fuel flow, and fuel

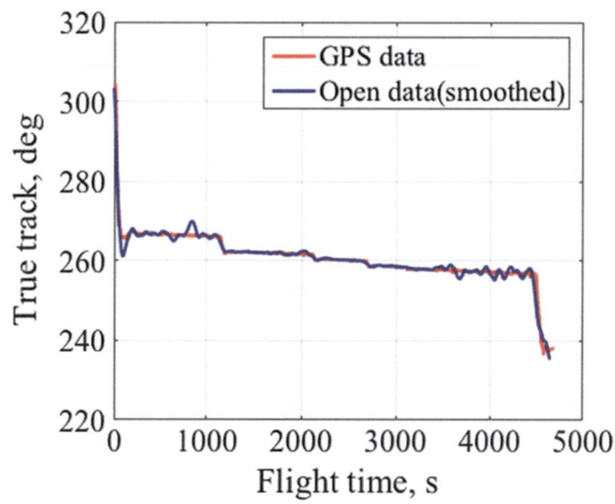

Fig. 4a Comparison of performance parameters. True track angle

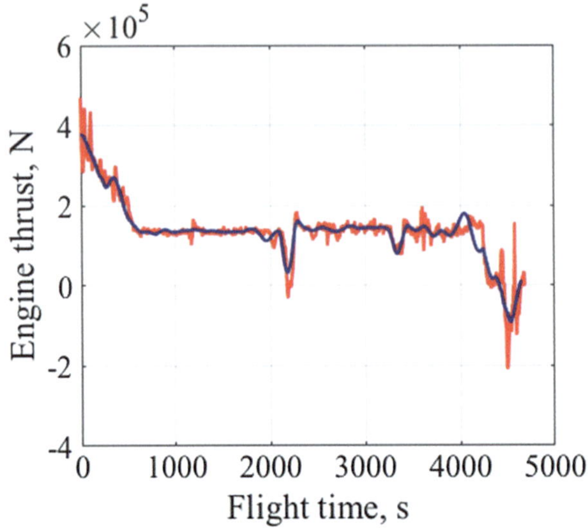

Fig. 4b Comparison of performance parameters. Engine thrust

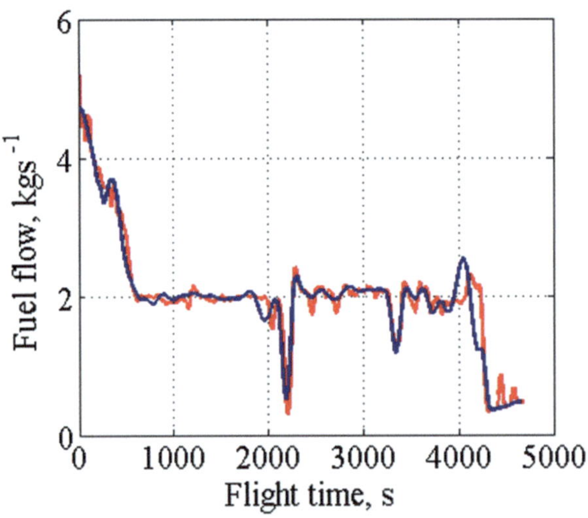

Fig. 4c Comparison of performance parameters. Fuel flow

consumption, with respect to flight time. Red line derives the results from GPS data, and blue line derives results from the smoothed open data. Results show that performance parameters from both sources are very much similar. This proves that the utilized smoothing process in this study provides data with considerable accuracy.

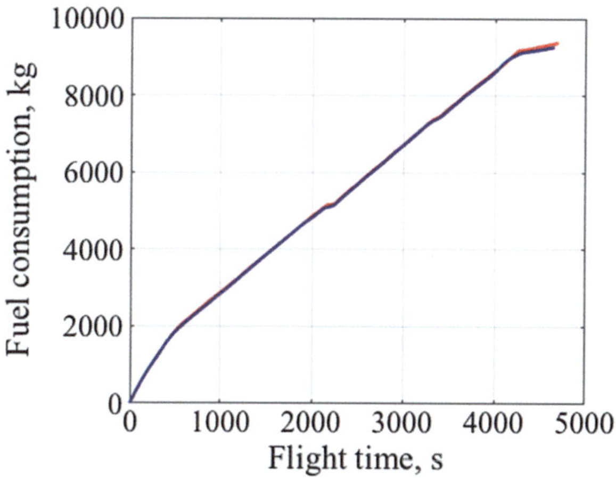

Fig. 4d Comparison of performance parameters. Fuel consumption

5 Operational Performance

The first part of the analytical results is explained here, mainly focusing on the performance of conventional ATC operations. Figures 5a, 5b, 6a and 6b resemble correlation characteristics between flight consumption and flight time and fuel consumption and flight time, respectively. To implement an impartial cost performance comparison, fuel consumption is normalized by the reference mass of each respective aircraft type provided in the BADA model. Solid lines denote the least-squares fittings (LSF) for each data series, and the black dotted line ($y = 0.00001x$) is given for reference.

Results show that the cost performance is inversely proportional to the scale of the aircraft. Also, the correlation between fuel consumption and flight time shows a stronger linearity than the correlation between fuel consumption and flight range. Pearson's correlation coefficients are calculated for each data series, and Table 4 provides the numerical values. Dispersions appear to be larger in (b) subplots compared to (a) subplot in both figures. This correlation shows that, furthermore, the dispersion is significantly larger in Fig. 6b compared to other plots, which has caused the lowest correlation coefficients compared to the performance in summer season. This is considered due to the existence of strong jet stream wind conditions over the Japanese airspace during the winter season. Depending on whether the aircraft is experiencing tailwinds or headwinds, fuel consumption shows significantly different values for similar flight range. Evaluated results derive that conventional domestic operations are not optimized for wind conditions and result in significant dispersions in the operational performance. The latter part of the analysis is introduced in the next section, where the performance assessment is reviewed on

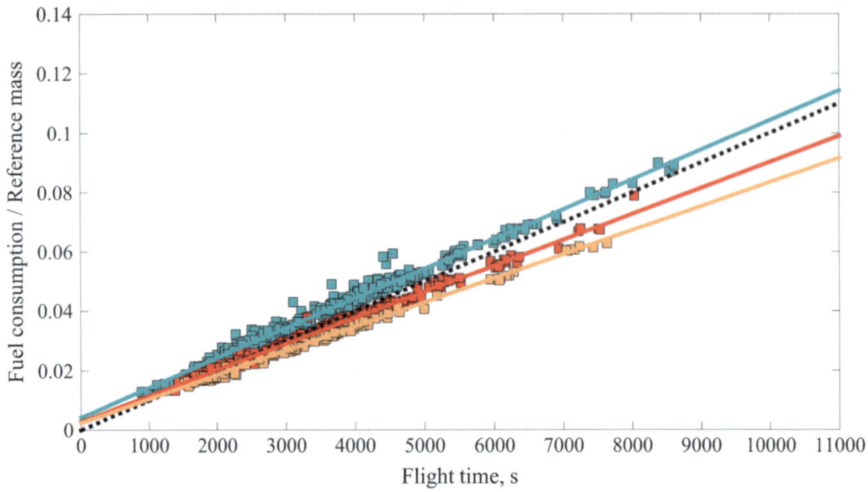

Fig. 5a Correlation characteristics between normalized fuel consumption and flight time. 2012/05 (summer season)

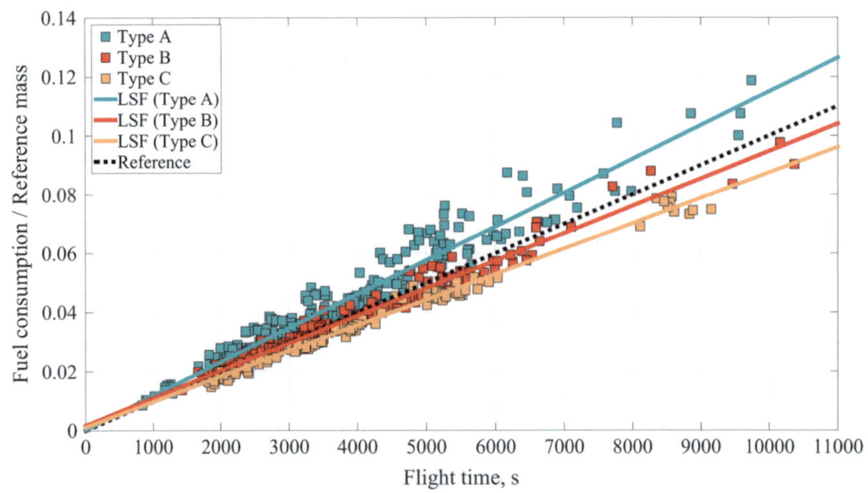

Fig. 5b Correlation characteristics between normalized fuel consumption and flight time. 2013/01 (winter season)

the assumption that the pilot is capable of performing ideal fuel-optimal flights by exerting the maximum performance of the aircraft with/without constraints on the flight time.

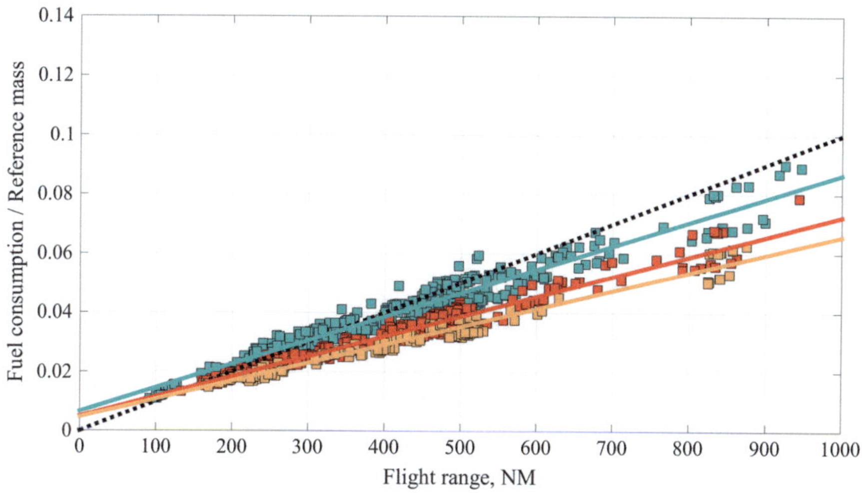

Fig. 6a Correlation characteristics between normalized fuel consumption and flight range. 2012/05 (summer season)

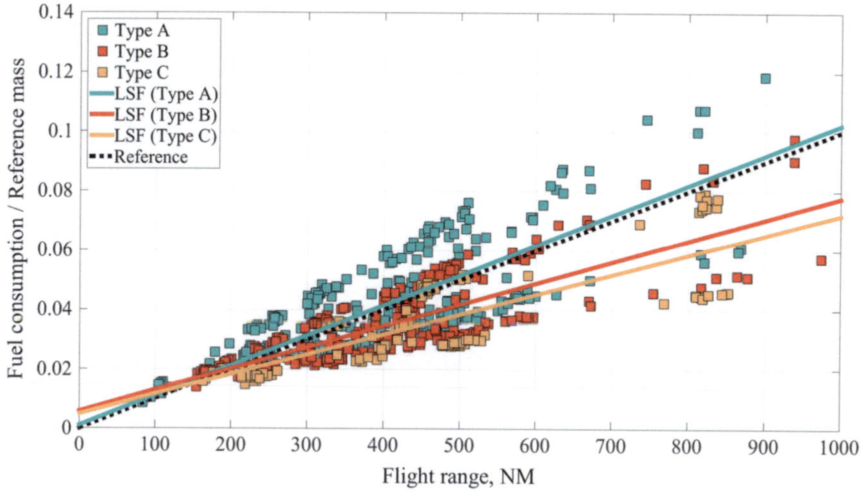

Fig. 6b Correlation characteristics between normalized fuel consumption and flight range. 2013/01 (winter season)

Table 4 Evaluations of operational performance and correlation coefficient

	Pearson's correlation coefficient					
	2012/05 (summer season)			2013/01 (winter season)		
Correlation	Type A	Type B	Type C	Type A	Type B	Type C
Fuel consumption vs. flight time	0.9927	0.9950	0.9977	0.9723	0.9863	0.9930
Fuel consumption vs. flight range	0.9687	0.9748	0.9792	0.7629	0.8058	0.8135

6 Potential Benefit Assessment

A two-approach method is used to discuss the potential benefits obtained through the application of 4D TBO concept. First, the fuel-optimal trajectories are generated without any penalty on the arrival time at the assigned destination. This method provides the maximum amount of potential benefits through 4D TBO compared to the conventional system. Previous studies show that this causes variations in flight time compared to the actual flight time of the corresponding flight. In the second approach, a CI value is assigned to the optimizer as a constraint on the arrival time. In other means, depending on the flight time difference between optimal and actual flights, a negative or positive penalty on the flight time is allocated to the performance index as the following:

- Optimal flight time – actual flight time < 0
  $\rightarrow$ penalty on flight time < 0 (increases flight time)
- Optimal flight time – actual flight time > 0
  $\rightarrow$ penalty on flight time > 0 (decreases flight time)

This enables to simulate fuel-optimal flight trajectories under conventional flight time constraints and possible time extending assignments in a future ATM system. Obtained results are used to discuss the deviation of potential benefits from the results obtained in the first approach.

Quantitative results for fuel-optimal trajectories without arrival time assignment are illustrated in Figs. 7a, 7b, 8a and 8b, and the numerical statistics are given in Table 5.

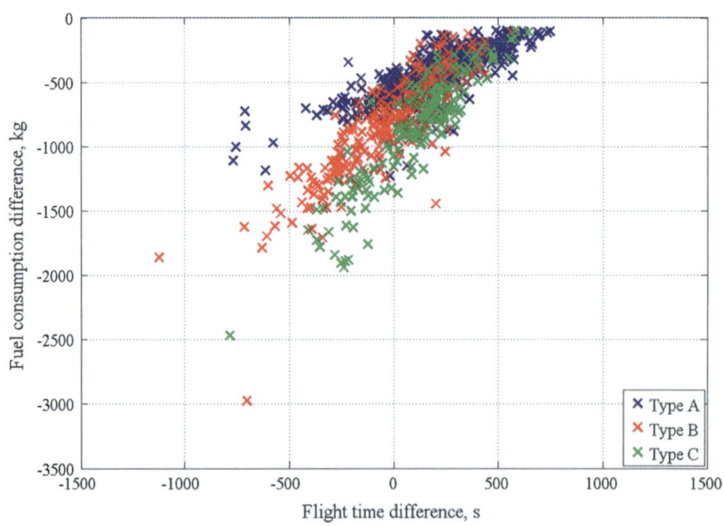

Fig. 7a Fuel consumption difference vs. flight time difference (without arrival time assignment). 2012/05 (summer season)

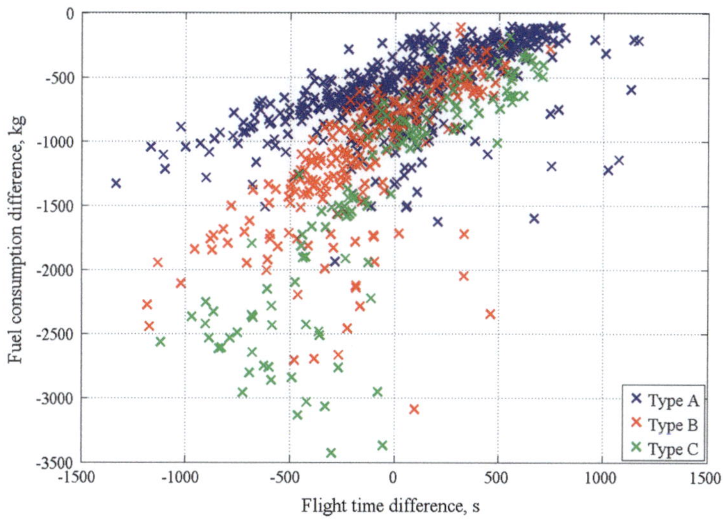

Fig. 7b Fuel consumption difference vs. flight time difference (without arrival time assignment). 2013/01 (winter season)

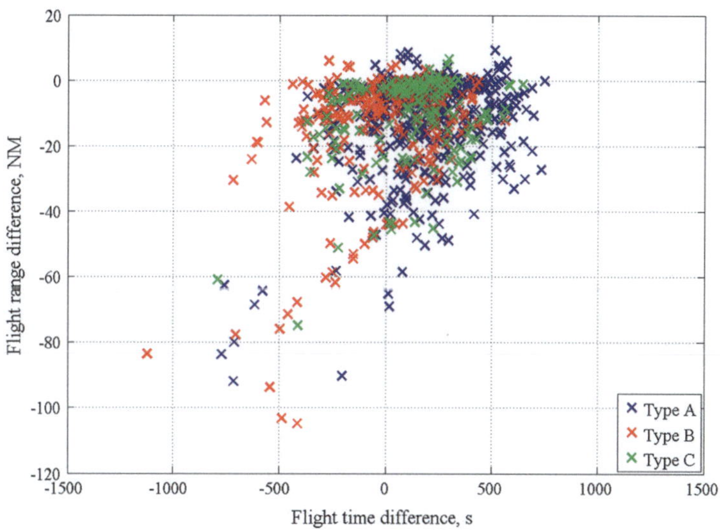

Fig. 8a Flight range difference vs. flight time difference (without arrival time assignment). 2012/05 (summer season)

Figures 7a and 7b denote the fuel consumption difference, and Figs. 8a and 8b denote the flight range difference for the subjected days in May 2012 and January 2013, respectively. All figures are plotted with respect to flight time difference.

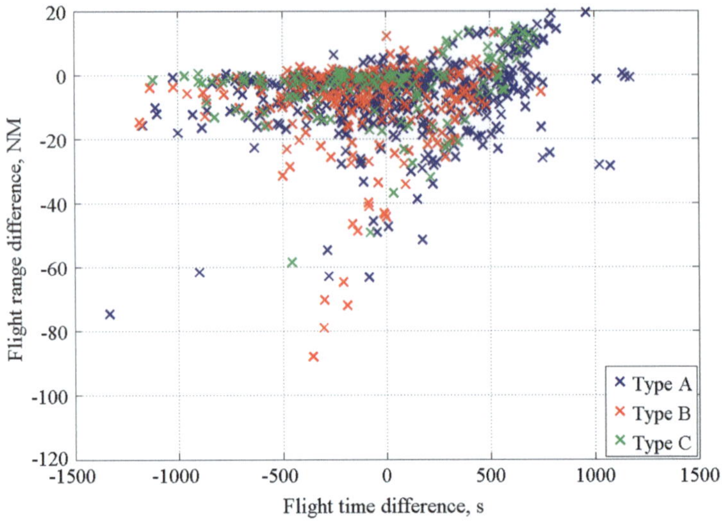

Fig. 8b Flight range difference vs. flight time difference (without arrival time assignment). 2013/01 (winter season)

Table 5 Potential benefits for optimal trajectories without arrival time assignment (average)

	Fuel consumption difference [kg]			Flight range difference [NM]			Flight time difference [s]		
	Type A	Type B	Type C	Type A	Type B	Type C	Type A	Type B	Type C
2012/05	−388	−756	−866	−11.5	−13.5	−10.7	236	−6	108
2013/01	−443	−922	−1167	−5.6	−8.4	−3.6	209	−25	40

These values are evaluated by subtracting the performance parameter of each actual flight from the corresponding performance parameter of optimal flight.

Results show that the trajectory optimizer is enabled to reduce the fuel consumption in both cases. Furthermore, by comparing the numerical values provided in Table 5, it is understood that the reduction of fuel consumption for January 2013 is greater than May 2012. This is achieved by trading off with exceeding flight time and shortening flight range compared to actual flights. Comparing the statistics, flight range and flight time differences between the two subjected days show contrary characteristics. It is considered that this phenomenon occurs mainly due to the influence of weather conditions. Fuel consumption reduction shows a linear trend with flight time trade-off for May 2012 compared to the largely dispersed results in case of January 2013. Also, compared to results in May 2012, there are more flight cases in January 2013 which show positive flight range difference against actual flights. This is because optimal trajectories in January 2013 show a greater tendency to use detours that ensure maximum effect of wind conditions compared to the selection of great circle route by most of the optimal trajectories in May 2012.

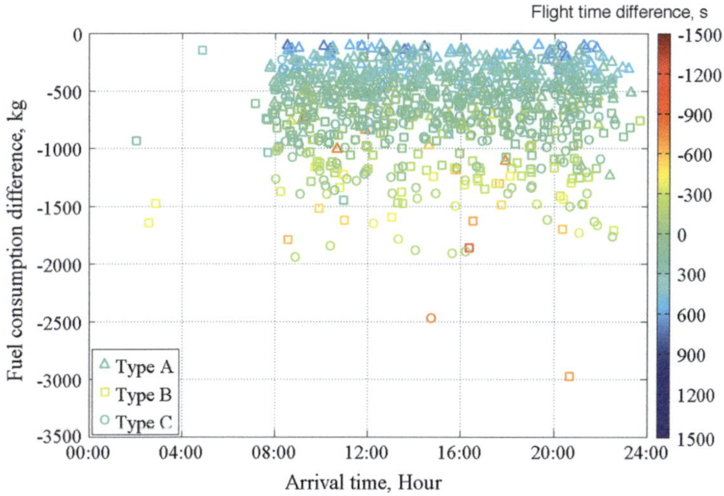

Fig. 9a Fuel consumption difference vs. arrival time (color bar: flight time difference). 2012/05 (summer season)

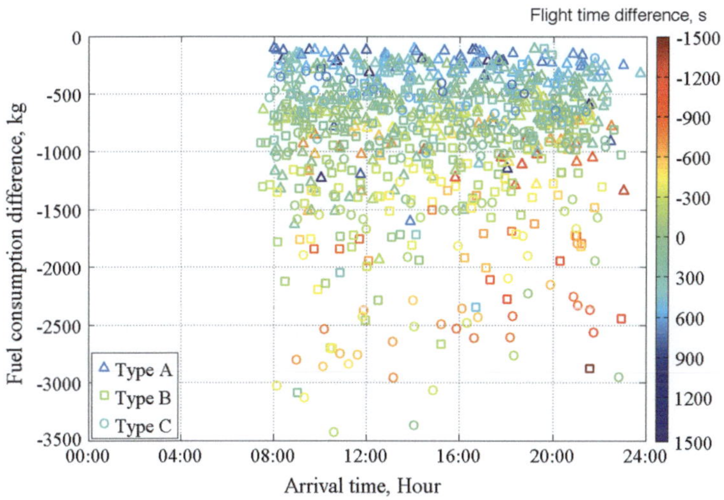

Fig. 9b Fuel consumption difference vs. arrival time (color bar: flight time difference). 2013/01 (winter season)

The fuel consumption reduction is plotted with respect to arrival time of the actual flight in Figs. 9a and 9b. The color distribution of each plot indicates the corresponding flight time difference. Results are plotted as Type A in triangular markers, Type B in square markers, and Type C in circular markers. Results in both cases reveal that fuel reduction does not have any propensity with the rush hour time periods. In Fig. 9b, significant fuel reduction is obtained throughout the subjected time period. This denotes that traffic congestion during rush hour is not the sole cause of less efficient operations. This leads to a consideration that air traffic handling methods used in conventional operations are also highly possible causes in such aspect. Hence, these results emphasize that user-prioritized time-based operations would have a significant impact on a future ATM system upon the conventional airspace-based operational system.

The second approach is considered in the following analysis. According to commercial pilots, CI is usually set to 80 in the flight management computer (FMC) for Type C aircraft. This corresponds to $a = 1$ in the defined performance index. Hence, for Type A and B aircraft, penalty values are allocated as a ratio of Type C aircraft mass. Evaluated values are given in Table 6. Typically, FMS are set with positive value for CI, and flight operations with extended flight time than the original flight plan are performed by speed adjustment and thrust setting followed by vectoring procedures at the terminal area. Negative CIs in Table 6 indicate the expression of flight time extension in a 4D TBO-based ATM system.

Potential benefits are compared between optimized trajectories with free arrival time and with assigned arrival times. Figures 10a, 10b, 11a and 11b illustrate the comparisons of potential benefits between fuel consumption and flight time and

Table 6 Cost index setting for flight time constraint application

	Cost index (penalty on flight time)	
	Positive	Negative
Type A	25 (0.3)	−25 (0.3)
Type B	60 (0.7)	−60 (−0.7)
Type C	80 (1.0)	−80 (−1.0)

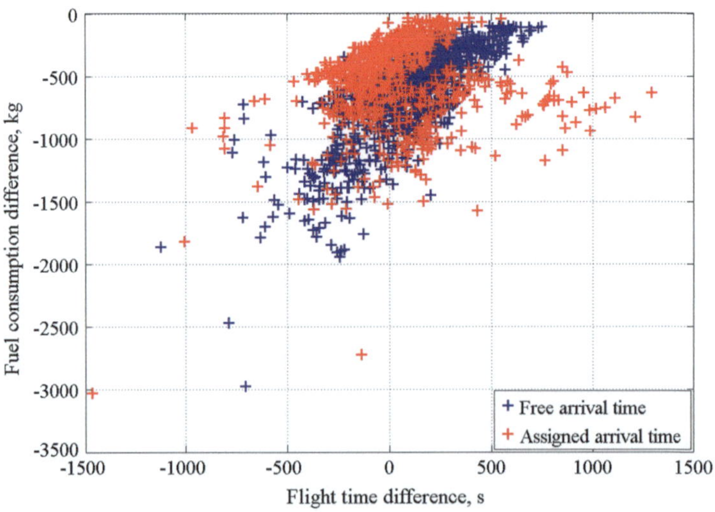

Fig. 10a Potential benefits comparison between fuel consumption and flight time. 2012/05 (summer season)

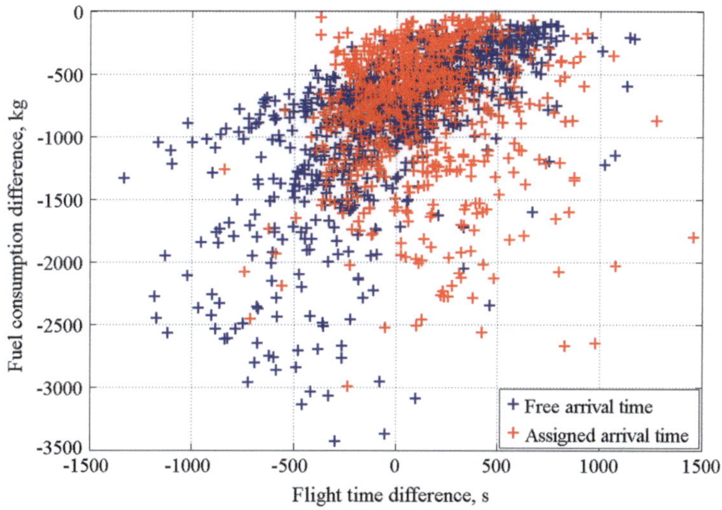

Fig. 10b Potential benefits comparison between fuel consumption and flight time. 2013/01 (winter season)

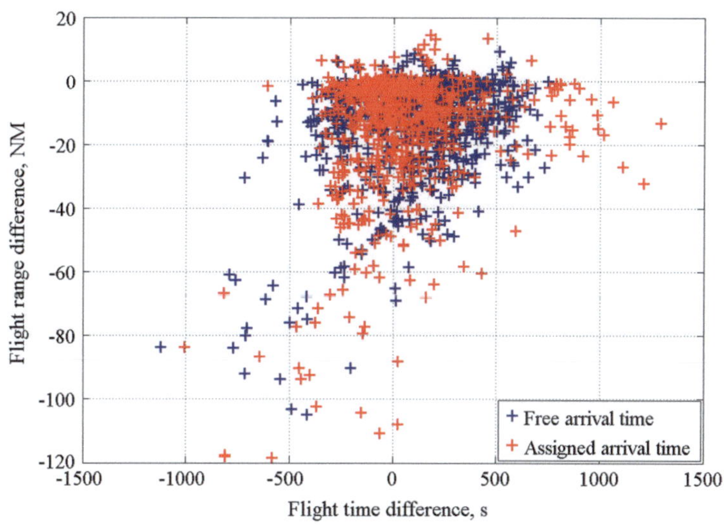

Fig. 11a Potential benefit comparison between flight range and flight time. 2012/05 (summer season)

flight range and flight time, respectively. Statistical expressions are given in Table 7. Table 8 shows the number of aircraft that is allocated to CI setting introduced in Table 6.

The blue color markers resemble the potential benefits from optimal trajectories with free arrival time, and red color markers resemble the potential benefits from

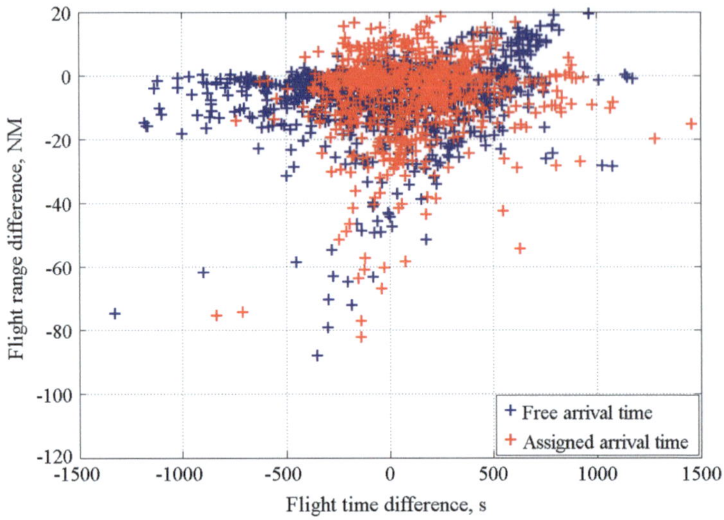

Fig. 11b Potential benefit comparison between flight range and flight time. 2013/01 (winter season)

Table 7 Potential benefits for optimal trajectories with arrival time assignment (average)

	Fuel consumption difference [kg]			Flight range difference [NM]			Flight time difference [s]		
	Type A	Type B	Type C	Type A	Type B	Type C	Type A	Type B	Type C
2012/05	−366	−678	−597	−16.1	−14.6	−10.6	21	18	154
2013/01	−388	−820	−872	−5.4	−7.3	−1.8	203	53	245

Table 8 Number of aircraft allocated for arrival time assignment

	2012/05			2013/01		
	Type A	Type B	Type C	Type A	Type B	Type C
Negative CI	72	135	45	173	154	72
Positive CI	375	155	124	313	119	81

optimal trajectories with assigned arrival time. Results in Figs. 10a and 10b reveal that the reduction of fuel consumption is decreased due to arrival time assignment. A part of the results in assigned arrival time scenario shows a significant extension of flight time in Figs. 10a and 10b. These are the optimized results for Type C aircraft with CI at −80. Hence, it is understood that CI setting for flight time extension causes a large penalty on the performance index. In other means, results emphasize that flight time extension can be achieved with a fraction of adjustment in CI setting without inflicting a significant negative impact on obtained benefits.

Results also indicate that, in order to comply with this constraint, optimal trajectories are generated with the trade-off of flight time and flight range. Authors' previous studies show that optimal trajectories tend to adjust its top of descent

(TOD) setting and descent speed in order to comply with the applied constraints. Also, it is understood that optimal trajectories without arrival time assignment perform a long descent phase with low-speed selection to maximize the fuel reduction, which extends the flight time. Hence, the variations in potential benefits in Figs. 10a, 10b, 11a and 11b indicate that optimal trajectories can compensate various constraints in the system by adjusting speed selection and TOD setting rather than following vectoring procedures in conventional operations. As a result, the ability to consider weather conditions in generating optimal paths, ability to exert the maximum performance possible from the aircraft, and ability to implement a user-prioritized system have enabled the achievement of expected benefits considering the entire operational system. These results strengthen the fact that 4D TBO has a significant impact on achieving an efficient ATM system in the future and could be a potential solution to meet the ever-increasing demands in the aviation industry.

7 Conclusion

This research study is focused on the performance assessment of 4D TBO based on a series of radar track data, publicly distributed under the CARATS Open Data program by the Japan Civil Aviation Bureau. A trajectory optimization method developed by the authors is applied to generate fuel-optimal trajectories with and without arrival time assignment to distinguish potential benefits that can be obtained by providing the pilot to perform efficient flight missions through the exertion of maximum performance of the aircraft. Three aircraft types operated in domestic flight operations are subjected to two different seasonal conditions, and trade-off between the performance parameters of fuel consumption, flight time, and flight range is discussed. Quantitative results show that weather conditions have a significant impact on potential benefit estimation. Optimal trajectories without time assignment tend to reduce fuel consumption by trading off with exceeding flight time and exceeding/shortening flight range according to weather conditions. The arrival time assignment reduces the overall fuel consumption reduction in both subjected days but address conventional operational restrictions while providing fuel savings. Quantitative results emphasize the fact that 4D TBO has a significant impact achieving a more efficient ATM system while meeting the increasing demands in the aviation industry.

The study is expected to enhance its scope by broadening the inclusion of other operational constraints, such as airspace restrictions and operational costs besides fuel and time costs, in the optimizer to understand the behavior of optimal trajectories and variations in potential benefits. Also, critical areas are to be declared in achieving such a system [21], and solutions are to be reviewed to obtain more realistic results and contribute toward the implementation of 4D TBO.

Acknowledgments Authors acknowledge the assistance of Prof. Yoshikazu Miyazawa of Kyushu University for providing the GPS-recorded flight data and Mr. Ralph E. Bach of Aerospace Computing, Inc., and Mr. Russell A. Paielli of NASA Ames Research Center for providing technical assistance with the smoothing algorithm.

References

1. Japan Civil Aviation Bureau (2010) Long-term vision for the future of air traffic systems (Changes to intelligent air traffic systems)
2. Oka M, Fukuda Y, Uejima K (2014) Outline of air traffic open data and application. In: 52nd aircraft symposium, Nagasaki
3. Brown M, Fujita M, Fukuda Y, Hirabayashi H, Inoue S, Nagaoka S (2013) Full 4D trajectory based operations concept study. In: 2013 Asia-pacific international symposium on aerospace technology (APISAT 2013), Takamatsu
4. Bourgois M, Sfyroeras M (2014) Open data for air transportation research: dream or reality. In: 14th international symposium on open collaboration (OpenSym'14)
5. Dobruszkes F (2012) Stimulating or frustrating research? Transport geography and (un)available data. Belgian J Geo(inaugural)
6. Good DH, Roller L-H, Sickles RC (1995) Airline efficiency differences between europe and the US: Implications for the pace of EC integration and domestic regulation. Euro J Oper Res (ELSEVIER) 80(3):508–518
7. Martin JC, Roman C (2001) An application of DEA to measure the efficiency of spanish airports prior to privatization. J Air Tran Man (ELSEVIER) 7(3):149–157
8. Bazargan M, Vasigh B (2003) Size versus efficiency: A case study of US commercial airports. J Air Tran Man (ELSEVIER) 9(3):187–193
9. Bhadra D (2009) Race to the bottom or swimming upstream: Performance analysis of US airlines. J Air Tran Man (ELSEVIER) 15(5):227–235
10. Wickramasinghe NK, Miyamoto Y, Harada A, Kozuka T, Shigetomi S, Miyazawa Y, Brown M, Fukuda Y (2014) Flight trajectory optimization for operational performance analysis of jet passenger aircraft. Aero Tech Japan Trans Japan Soc Aero Space Sci 12(APISAT-2013): a17-a25
11. Harada A, Kozuka T, Miyazawa Y, Wickramasinghe NK, Brown M, Fukuda Y (2014) Analysis of air traffic efficiency using dynamic programming trajectory optimization. In: 29th Congress of the International Council of the Aeronautical Sciences (ICAS 2014)
12. Miyazawa Y, Matsuda H, Shigetomi S, Harada A, Kozuka T, Wickramasinghe NK, Brown M, Fukuda Y (2015) Potential benefits of arrival time assignment – Dynamic programming trajectory optimization applied to the tokyo international airport. In: 11th USA/Europe Air Traffic Management Research and Development Seminar (ATM 2015)
13. Wickramasinghe NK, Harada A, Totoki H, Miyamoto Y, Miyazawa Y (2014) Flight trajectory optimization for modern jet passenger aircraft with dynamic programming. Air Traf Man Sys Lect Notes Elec Eng 290:87–104
14. Wickramasinghe NK, Brown M, Fukushima S, Fukuda Y, Harada A, Miyazawa Y (2015) Correlation between flight time and fuel consumption in airliner flight plan with trajectory optimization. In: AIAA Guidance, Navigation and Control Conference (SCITECH 2015), Florida

15. Fukuda Y, Oka M, Wickramasinghe NK, Uejima K (2015) Air traffic open data and its improvement. In: 2015 Asia-Pacific International Symposium on Aerospace Technology (APISAT 2015), Cairns
16. Japan Aeronautical Information Service Center (2015). https://aisjapan.mlit.go.jp/Login.do
17. Japan Meteorological Business Support Center Online Data Service (2006) (in Japanese). http://www.jmbsc.or.jp/hp/online/fonline0a.html
18. Eurocontrol Experiment Center (2011) User manual for the Base of Aircraft Data (BADA). Revision 3.11, EEC Technical/Scientific Report, No. 13/04/16–01
19. Roberson B (2007) Fuel conservation strategies, cost index explained. Aero Quar QTR 02–07
20. Bach RE, Paielli RA (2014) A user guide for smoothing air traffic radar data. NASA/TM-2014-216520
21. Nagaoka S, Brown M (2015) Constructing an index of difficulty for air traffic control using proximity parameters. J Proc Eng 99:253–258

Future Arrival Management Collaborating with Trajectory-Based Operations

E. Itoh, M. Brown, A. Senoguchi, N. Wickramasinghe, and S. Fukushima

Abstract The International Civil Aviation Organization (ICAO) has created a long-term plan for a harmonized global future air traffic management (ATM) system to be achieved by the year 2030. A key element of this plan is four-dimensional trajectory-based operations (4D TBO), where aircraft fly along optimal trajectories defined in space and time and agreed to between the aircraft and air traffic controllers and operators. One issue is how to minimize the uncertainty of predicted arrival time, which increases in proportion to flight distance from the destination airport. To clarify and help resolve this issue, this study investigates design principles and algorithms for a novel ground advisory system which smoothes arrival air traffic by providing coverage for both en route and terminal airspaces. Defined as "Extended Arrival MANager (E-AMAN)," operational concepts to be used to collaborate with 4D TBO are proposed in this paper. Information sharing, air-ground harmonization, and the design of human-system interactions are discussed as three main technologies for supporting efficient arrival operations in the future. Furthermore, future arrival scheduling should not only follow the current first-come first-served (FCFS) protocol but should also consider performance-based operations (PBO) targeting the mixed equipage situation in a future ATM environment. This paper presents a future vision of 4D TBO-based arrival management and clarifies policies for developing technologies to support the ATM system of 2030.

Keywords Extended Arrival Management (E-AMAN) • 4D trajectory-based operations (4D TBO) • Performance-based operation (PBO) • System Wide Information Management (SWIM) • Air-ground harmonization • Human-computer interaction (HCI)

E. Itoh (✉) • M. Brown • A. Senoguchi • N. Wickramasinghe • S. Fukushima
Air Traffic Management Department, Electronic Navigation Research Institute (ENRI), Tokyo, Japan
e-mail: eri@enri.go.jp

Electronic Navigation Research Institute, *Air Traffic Management and Systems II*, Lecture Notes in Electrical Engineering 420, DOI 10.1007/978-4-431-56423-2_7

1 Introduction

ICAO draws a future vision of air traffic management (ATM) as moving from current airspace-based operations to trajectory-based operations (TBO), which manage all aircraft trajectories from departure to arrival [1]. The ultimate realization of TBO is called full four-dimensional TBO (full 4D TBO), in which aircraft fly to the maximum possible extent on user-preferred trajectories defined by time (airspeed), latitude, longitude, and altitude. If Full 4D TBO is applied to all air traffic, all aircraft will be able to arrive at their destination on time while minimizing operational costs. The infrastructure, procedures, and regulations to enable the 4D TBO concept are planned to be realized around the year 2030. However, there are several technical issues to overcome before introducing the 4D TBO in the future ATM environment.

One question to clarify is how to minimize the impact of uncertainties in flight, which increases prediction errors of aircraft arrival time in proportion to flight distance from the destination airport. Uncertainty in flight can arise from weather changes, human factors in operation (e.g., pilot and air traffic controller (ATCo)), and difference in airborne equipage including flight management system (FMS) functions and performance. A variety of airborne equipages will lead to the ATM system having to accommodate aircraft with a range of performance-based operation (PBO) capabilities in the year 2030. Not only the conventional first-come first-served (FCFS) protocol but also the best-equipped best-served (BEBS) protocol will be required for fair arrival sequencing in the future PBO. To help resolve this issue, this study investigates design principles and algorithms for a novel ground advisory system defined as "Extended Arrival MANager (E-AMAN)" which provides coverage for both en route and terminal airspaces. This paper presents a future vision and operational concept of the E-AMAN to collaborate with 4D TBO targeting the projected arrival air traffic within the next two decades. Information sharing, air-ground harmonization, and the design of human-system interactions are clarified as three key technologies in this paper to support future arrival management.

This paper is organized as follows. Section 2 summarizes concepts and preceding studies of the 4D TBO and ground advisory system for arrival management. Section 3 presents a future vision of designing the 4D TBO-based E-AMAN. Operational concepts are proposed to harmonize future arrival management with 4D TBO. Section 4 provides a policy for developing technologies which realize the proposed E-AMAN operational concepts. All required information for arrival operations should be transferred to the ground-based E-AMAN. These information sets should be shared among air-ground systems and be structuralized for their efficient use in a distributed computing platform. Technologies to support future 4D TBO-based arrival management are summarized in this section. Section 5 concludes this paper and discusses our future works.

2 4D TBO and Ground Advisory System for Arrival Management

2.1 4D TBO

4D TBO is an operations concept which manages all aircraft 4D trajectories (4DTs) defined in space and time from departure to arrival. The 4DTs planned to be flown by an aircraft are negotiated and agreed to between the pilot, ATCos, and airline operators. The details and uncertainty of planned trajectory information depend on the planning of or updating of the time scale. In time scales from days in advance of the departure to in-flight operation, the 4DTs are progressively refined and updated to balance demands with capacity under negotiations between all stakeholders, based on the so-called collaborative decision-making (CDM). 4DT information is shared between all involved stakeholders via a data interchange network known as System Wide Information Management (SWIM). Information related to each flight including planned trajectory information, aircraft type, and capabilities information is bundled as a "flight object" in the SWIM system and is accessible by stakeholders in real time. During flight, each aircraft's future 4DTs are estimated by its on-board FMS and by ground-based systems independently. Since these "airborne" and "ground" estimated trajectories are computed with different algorithms from different data with different qualities, discrepancies between the two will arise. In order to synchronize airborne and ground trajectories, the ground system will receive copies of aircraft's FMS computed 4DT via datalink. While ground systems can compute estimated trajectories for different weather scenarios and possible routes, the aircraft FMS generally only computes the trajectory on the programmed flight path.

In the authors' other works at ENRI, the benefits of introducing full 4D TBO are being assessed targeting predicted Japanese air traffic flows in the year 2030 [2]. For designing 4DT while avoiding congestion in Japanese airspace, an index of airspace complexity is being developed to support efficient trajectory management [3]. A method of trajectory optimization is also being developed to design operator-preferred 4DTs that take forecast winds into account and achieve a given trade-off between fuel efficiency and flight time to meet flight schedules with minimum operational cost [4, 5]. Figure 1 shows an example of optimized 4DT routes arriving at the Tokyo International Airport.

4D TBO-based energy-saving arrivals have also been studied by the authors [6–8]. One of the concepts for realizing energy-saving arrivals is Continuous Descent Operation (CDO) [9], in which aircraft descends continuously from cruise to the airport (i.e., without intermediate level offs) at near-idle thrust. Since unpredictable individual aircraft trajectories during CDO leads ATCo to increase separations for safety reasons, CDO is currently only applicable to low-density air traffic. In order to apply CDO to high-density traffic by improving trajectory predictability, the authors have proposed to apply a "fixed-flight-path angle (FPA) descent" concept in which arrival aircraft continuously descend to the assigned runway following the fixed vertical path angles as shown in Fig. 2. The operational feasibility of FPA

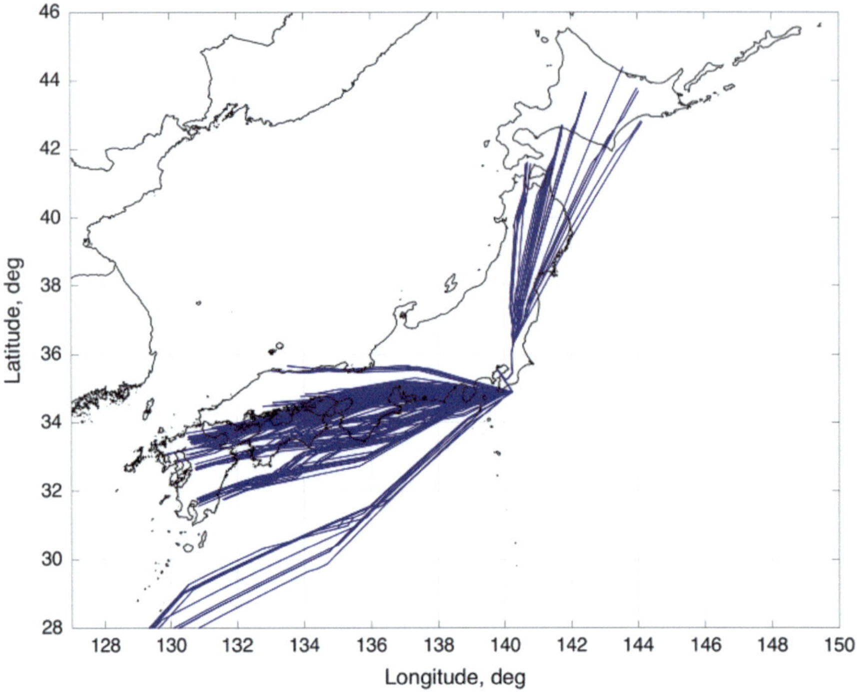

Fig. 1 An example of optimized 4DT for arriving at the Tokyo International Airport

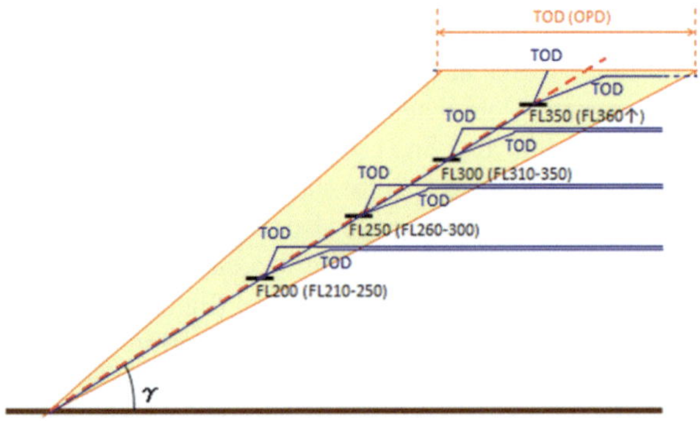

Fig. 2 Concepts of the FPA descent. *In this example, four merging points are set at FL200, FL250, FL300, and FL350: arrivals which cruise above FL360 merge to the FPA descent path at FL350, above FL310 below FL360 merge at FL300, above FL260 below FL310 merge at FL250, and above FL210 below FL260 merge at FL200. FPA descent does not specify the position/altitude of the top of descent (TOD) of arrival aircraft. Due to the TOD altitude, the FPA descent provides options of the altitude where the arriving aircraft start their continuous descent. The values of FPA should be selected in order to minimize fuel consumption*

Fig. 3 B777-200 full flight simulator

descents was evaluated using an airline's B777–200 full flight simulator as shown in Fig. 3. Flight deck Interval Management (FIM), which is one of the applications of the Aircraft Surveillance Applications System (ASAS), was applied to control airspeed during FPA descent [6]. These studies indicate that FPA descent for efficient arrival management is feasible.

2.2 *Ground Advisory System for Arrival Management*

Conventionally, ground advisory tools which support ATCos were established to realize efficient arrival scheduling by automatically calculating arrival sequences and scheduled times of arrival (STA), so-called Arrival MANager (AMAN) in Europe. In the United States, Traffic Management Advisor (TMA) was introduced to ground centers in the 1990s to smooth arrival operations from en route to terminal airspace. Since then, TMA has been continuously upgraded for almost two decades: the latest version is called TBFM (Time-Based Flow Management) for en route areas and TSAS (Terminal Sequencing and Spacing) for terminal areas. TMA consists of three components: route analyzer (RA), traffic synthesizer (TS), and dynamic planner (DP) [10, 11]. The RA calculates aircraft trajectories from current positions to arrival airports following given flight plans and ATCo interventions such as speed instructions and radar vectors. The TS calculates estimated time of arrival (ETA) at terminal airspace entry points (called *terminal gates*) and the planned runway threshold at the airport following the RA-computed trajectories. The TS also calculates ETA at alternative runways at the airport when the aircraft arrive at the airport with multiple runways. The DP generates an arrival schedule, which determines arrival sequences and scheduled times of arrival (STA) at *terminal gates* and runway thresholds using the ETA values estimated by the TS. The DP can also propose changing the arrival runway to minimize runway slot losses. Runway allocation following the DP proposal triggers a new trajectory to be computed by the RA.

In this way, the DP collaborates with the TS and RA for arrival scheduling. One of the authors' references [12] explains the algorithms of arrival scheduling following FCFS protocols. In the future, not only the FCFS protocol but also a BEBS protocol will be required to achieve fair arrival schedules taking into account the advanced aircraft capabilities and performance. Various advanced operational procedures will be gradually increased in future air traffic corresponding to a variety of airborne equipages. Future ground systems will be required to support arrival scheduling in an anticipated mixed equipage situation. Examples of the mixed equipage air traffic will be discussed in Sect. 4. In the practical use of the TMA system, the frequency and timing of arrival schedule updates by the DP is one of the key issues that affect performance. The RA updates aircraft trajectories when triggered by flight plan changes or ATCo interventions. The TS updates ETA every 10–15 s considering the current radar update cycle. The DP updates the arrival schedule when an arrival aircraft crosses a "freeze horizon," which lies in en route airspace 15–20 min before the arrival aircraft is predicted to cross the *terminal gate*.

3 Operational Concepts of E-AMAN Collaborating with 4D TBO

3.1 Environmental Assumptions

This section proposes the operational concepts of the future E-AMAN system targeting arrival traffic to the Tokyo International (Haneda) Airport, starting at approximately 100–150 NM from the runway threshold. Arrival air traffic enters terminal airspace from multiple directions and arrives at multiple runways. It is assumed that all ground-based infrastructures and airborne systems required for conducting 4D TBO are implemented, including automation, air-ground datalink, and human-machine interfaces, and these operational procedures are established. Planned 4DTs, which are determined by agreements between airlines and air navigation service providers, are updated by both ground and airborne systems. ATCos and pilots communicate primarily by datalink to conduct 4D TBO. Pilots control and navigate aircraft to follow the negotiated 4DT.

It is assumed that the negotiated 4DT is transferred to the aircraft using the SWIM network. The E-AMAN uses the most recent 4DT by retrieving it from the aircraft's flight objects, since it is supposed that the flight object copy will reflect flight plan changes and the results of ATCo intervention and be automatically synchronized with the FMS trajectory if a significant discrepancy arises between the ground and airborne estimates.

We assume arrival traffic to have "mixed equipage," with different levels of guidance system (FMS) and RNP capability and ASAS equipage. More details on advanced operations in mixed equipage air traffic will be discussed in Sect. 4.

3.2 Functions in the Future E-AMAN

Based on the three functions of TMA explained in Sect. 2.2, we propose counterpart functions for the future E-AMAN that accommodate the capabilities of 4D TBO, namely, 4DT-RA, 4DT-TS, and 4DT-DP. Figure 4 shows the outputs and interactions among these three functions and with external systems.

The 4DT-RA estimates and updates the *4DT analyzed route* of arrival aircraft which enters into the airspace covered by E-AMAN (*E-AMAN airspace*). *4DT analyzed route* is the arrival route from the aircraft's current position to the runway threshold following the 4DT planned in the flight object. We assume that the arrival runway and *terminal gate* are initially assigned before the aircraft enters *E-AMAN airspace*. Once the amended trajectory has been negotiated with the aircraft, the planned trajectory of the aircraft's flight object will be updated, and the 4DT-RA will then receive the agreed updated trajectory via SWIM. So, the 4DT in the flight object transferred to the 4DT-RA is the one which is agreed upon among airlines and service providers and saved into the flight objects by the external ground system for 4D TBO. 4DT-RA estimates and updates the *4DT analyzed route* in E-AMAN when the 4DT is updated in the flight objects. Additionally, 4DT-RA updates the *4DT analyzed route* when 4DT-DP and/or ATCo changes arrival runways for efficient arrival scheduling. When the 4DT-RA computes the *4DT analyzed route*, ETA to

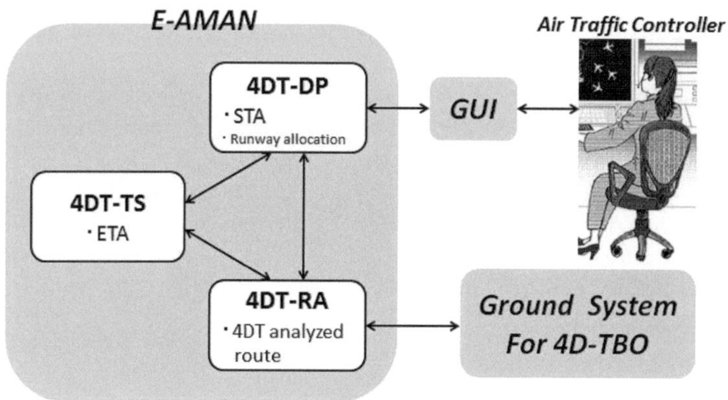

Fig. 4 Functions and interactions of the proposed E-AMAN

the assigned waypoints and runway threshold following the planned 4DT are able to be obtained by calculating flight time following the lateral and vertical paths on the *4DT analyzed route.*

The 4DT-TS estimates and updates the ETA of the arrival aircraft at assigned waypoints and the runway thresholds following the *4DT analyzed route*. 4DT-TS estimates and updates the ETA at least every 10–15 s, to improve the ETA prediction accuracy. Additionally, the 4DT-TS estimates ETA on the 4DT to the alternative runways when the aircraft arrive at an airport with multiple runways. These ETAs are also updated at least every 10–15 s. In the cases where we assume the advanced radar technology (e.g., SSR Mode S 1090 ES) and/or ADS-B technology for the year 2030, high frequencies of the updates will be expected to improve ETA estimation accuracy in the 4DT-TS.

The 4DT-DP computes arrival schedules which consist of arrival sequences and STA at assigned points including runway thresholds, *terminal gates*, and merging points. The arrival schedules are calculated following the *4DT analyzed route* in the 4DT-RA and by using the values of ETA estimated by 4DT-TS. ATCos and pilots navigate the aircraft to fill the gap between STA and ETA. The 4DT-DP computes runway allocation for minimizing slot losses on the runway. In the case that the 4DT-DP allocates the arrival runway, the arrival schedule is computed targeting the newly assigned runway. The change of the arrival runway in the 4DT-DP is updated in the 4DT-RA and in the arrival information in the flight object.

For practical use in arrival management, the timing of updating arrival schedules in the 4DT-DP is a key for synchronizing 4DT between E-AMAN and 4D TBO. The next section proposes operational concepts on updating arrival scheduling in order to integrate with 4D TBO.

3.3 Operational Concepts

This section proposes operational concepts for harmonizing future operation between E-AMAN and 4D TBO by synchronizing update timings in the 4DT-RA and 4DT-DP.

In our proposed operational concept, airspace up to approximately 30–40 NM around the airport is defined as the *terminal area*. The point where an arrival aircraft enters the terminal area is defined as a *terminal gate*. Two *freeze horizons* are established in en route airspace beyond the terminal: an *arrival scheduling freeze horizon* (A-FH) at approximately 15–20 min before the arrival aircraft crosses the *terminal gate* and a *4DT optimization horizon* (O-FH) at approximately 5–10 min before crossing the A-FH. These configurations are shown in Fig. 5.

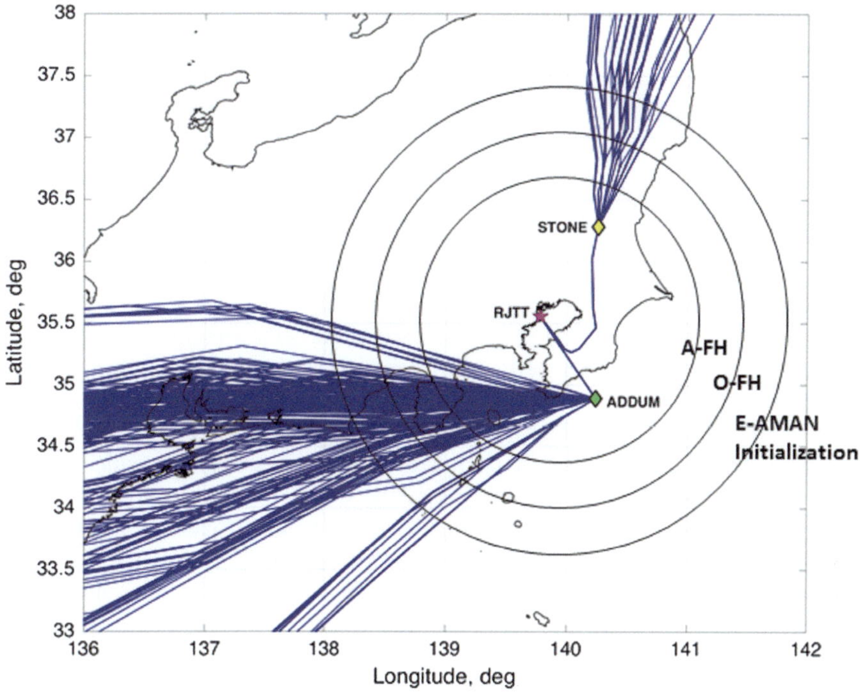

Fig. 5 Image of the operational concepts for the proposed E-AMAN: *In a precise sense, A-FH, O-FH, and E-AMAN initialization points do not allocate on the circle in this figure since these constraints are defined by time. Two terminal gates currently used for arrivals to the Tokyo International Airport, STONE and ADDUM, are assumed to draw the figure*

As shown in Fig. 5, arrival aircraft may follow arbitrary 4DTs agreed via CDM before crossing a *terminal gate*, but the paths from the *terminal gates* to the runway threshold are assumed to be based on fixed lateral routes with constrained vertical and speed profiles, similar to current standard terminal arrival (STAR) procedures. One of the fixed paths in the *terminal area* should be selected by CDM when the overall 4DT is designed. E-AMAN selects one of the fixed paths in the *terminal area* when it allocates arrival runways.

Arrival scheduling, which computes arrival sequences and STA, is triggered three times. The 4DT-DP determines the runway allocation, updates the arrival sequence, and assigns STA to the aircraft. The first is when an arrival aircraft enters the *E-AMAN airspace*. The second time is when the aircraft crosses the A-FH. The final update occurs when the aircraft crosses a *terminal gate*. By these repeated updates of the *arrival scheduling*, E-AMAN reduces the impacts of uncertainties in *arrival scheduling* which could cause delays. So to speak, the arrival schedule is "frozen" at least three times.

The frozen scheduled arrival sequence and STA may need to be revised due to situation changes (e.g., pop-up traffic, the need to reinsert a missed approach into the arrival sequence, weather changes, runway closure). We call this revision process *arrival rescheduling*.

Simultaneously with E-AMAN, part of the 4D TBO concepts works to balance demand and capacity similarly to current aircraft traffic management and flow management. Demand is predicted based on aircraft 4DTs, and if it exceeds capacity, trajectories are modified based on CDM between stakeholders. Such negotiated 4DT modifications in the flight objects should be updated before the arrival aircraft crosses the O-FH. Inside the O-FH, updates to the 4DTs should be carried out by the E-AMAN and/or ATCo for efficient air traffic control, for example, to enable short-term tactical interventions by ATCos and runway allocation.

Figure 6 shows a block diagram which explains the update timings of 4DT-RA and 4DT-DP. 4DT is updated by upstream trajectory management which should be done before the arrival aircraft crosses the O-FH. After crossing the O-FH, E-AMAN assumes responsibility for trajectory management, and aircraft 4DTs are updated due to *arrival scheduling* and *rescheduling* by 4DT-DP and/or ATCos. The 4DT-DP determines the runway allocation, updates the arrival sequence, and assigns

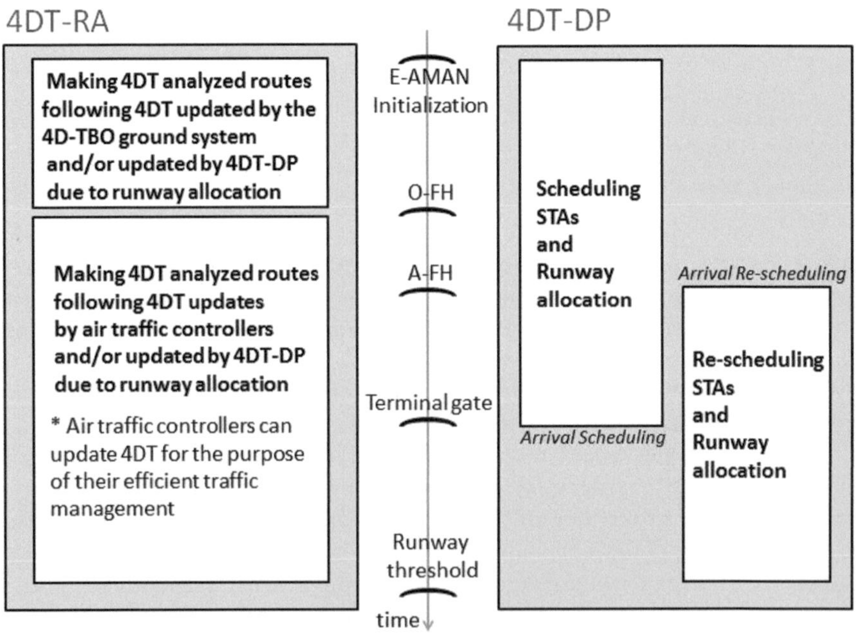

Fig. 6 Updates of the proposed E-AMAN in 4DT-RA and 4DT-DP

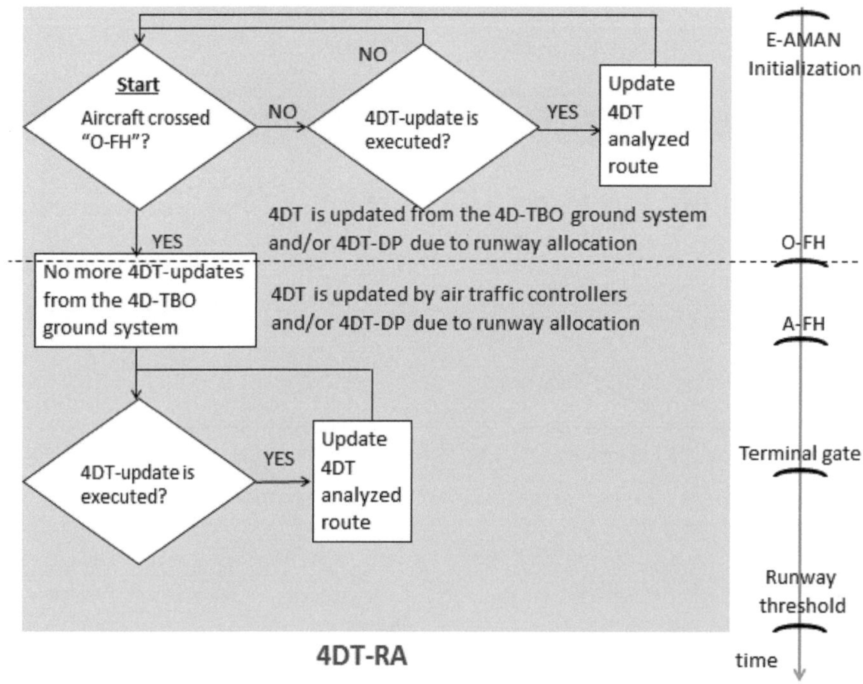

Fig. 7 Diagrams of updating processes in 4DT-RA

STA to the aircraft. The block diagrams in Fig. 7 show procedures for updating the 4DT-RA.

As shown in Fig. 6, 4DT-DP handles two tasks: updates of *arrival scheduling* and *arrival rescheduling*. Figure 8 shows procedures of these updating processes in 4DT-DP. As explained in the previous section, an aircraft's *arrival scheduling* is calculated and updated three times: when the E-AMAN is initialized, when the arrival aircraft crosses the A-FH, and when it crosses the *terminal gate*. *Rescheduling* is carried out after the arrival aircraft crosses the A-FH, if necessary. Strategic runway allocation is permitted to increase the efficiency of runway usage by minimizing slot losses. Arrival sequence considers BEBS protocol to exploit higher aircraft navigation and guidance system performance, not only FCFS protocols.

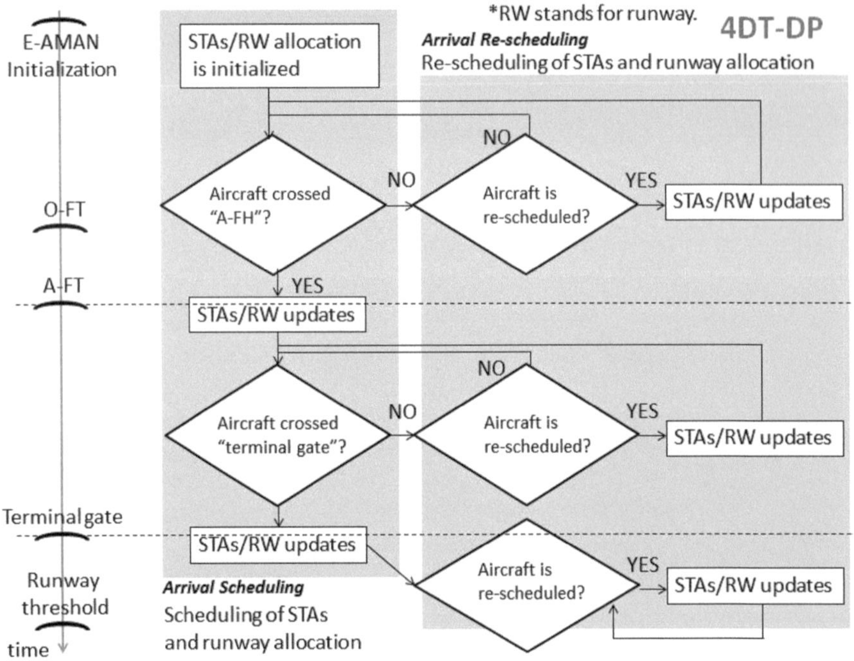

Fig. 8 Diagrams of the updates in 4DT-DP

4 Technologies to Support Future Arrival Management

4.1 Technical Challenges

This study focuses on the following three aspects, treated as future challenges for realizing the E-AMAN concept.

The first challenge is to develop the E-AMAN protocols. One of the distinctions between the proposed E-AMAN and conventional ground advisory system (e.g., AMAN, TMA) is their protocols in updating aircraft trajectories. The conventional systems generally operate as closed-loop systems since their designs assume that aircraft follows the scheduled flight path. The proposed E-AMAN is expected to synchronize with 4DT updates during flight. Furthermore, the protocols should be designed to be accepted by the ATCos and pilots during operation. This study reviews the suggested ideas in this paper and develops the proposed E-AMAN protocols.

The second challenge is to develop arrival scheduling algorithms in the E-AMAN. In the future traffic, various advanced operational procedures will gradually increase corresponding to the varieties of airborne equipages. For instance, one of the keys in 4D TBO is the datalink technology to share trajectories between air and ground. High-level RNP (RNP AR) is an example which enables curved approaches. It shortens flight time in the final approach phase while shifting arrival sequencing with other arrival aircraft. FIM and Advanced-IM (A-IM) are pair-based operations to keep self-separation using ADS-B information. Arrival sequences should not interfere with the aircraft pairs. The conventional arrival scheduling follows the FCFS protocol since it offers fair arrival sequencing to airlines and supports the tasks of ATCo who generally handle traffic on an FCFS basis. However, in order to obtain maximum benefits from the above mentioned airborne equipage capabilities, arrival scheduling is required to follow the BEBS protocol. This study investigates efficient arrival scheduling algorithms under the mixed equipage situation which is feasible for ATCos and pilots in terms of operation.

The third challenge is to design information flow models for air-ground, air-air, and ground-ground. To implement future E-AMAN, all required information should be shared as adequate structures among air-ground, air-air, and ground-ground. Precise models to show information networks will be clarified while designing protocols and scheduling algorithms.

In line with this, this paper picks up information regarding sharing and structuring, air-ground harmonization, and HCI design as three key technologies to support the above challenges. More details will be discussed in the following sections.

4.2 Information Sharing and Structuring

For implementing the future TBO-based E-AMAN proposed in this paper, all required information should be shared as adequate structures with the E-AMAN for computing efficient arrival schedules via SWIM networks.

Provisional information sets which are inputs in the proposed E-AMAN are picked up in Tables 1, 2, and 3. In addition to types of input information, their sources and update frequencies are summarized in the tables based on TMA in Refs. [10–12] and proposed E-AMAN operational concepts.

Table 1 is the list of information inputs to 4DT-DP, which generates arrival schedules. "Flight objects" is a package of the flight information including the 4DT. "Track data" includes current position and airspeed. "ETA" is updated in 4DT-DP. Constraints for STA calculation are summarized in the information inputs from number four to seven. "Separation time/distance" is the constraint between arrival aircraft at runway thresholds and assigned points. "Occupancy time" is the minimum amount of time between two landings at a runway threshold. "Acceptance rate" limits the number of aircraft crossing at an assigned point in a given time period. "Blocked interval" is the time interval when the runway and airspace are closed. These information sources should be shared in a common database for realizing the harmonization among arrival, departure, and surface traffic management. Constraints of the arrival sequencing are given in the eighth and ninth information inputs. "PBO information" defines the type of operation (e.g., ASAS and RNP AR) and its situation (e.g., executed or terminated). "ATCo's intention"

Table 1 Inputs to 4DT-DP

No	Input information	Information details	Sources	Update timing
1	Flight objects	Aircraft ID, types, characteristics, flight plans, updated 4DT, etc.	Center's host computer	4DT-DP is initialized/4DT is updated
2	Track data	Aircraft tracks in radar data	Center's host computer	All updates of 4DT-DP
3	ETA	ETAs (maximum, nominal, and minimum value of ETA) at runway threshold, terminal gate, and other assigned points	4DT-TS	All updates of 4DT-DP
No. 4–7 scheduling constraints				
4	Separation time/distance	Miles-in-trail, minimum horizontal/lateral separation, wake vortex separation, etc.	Center's host computer	4DT-DP is initialized/separation time/distance is updated
5	Occupancy time	Minimum time interval between landings	Center's host computer	4DT-DP is initialized/occupancy time is updated
6	Acceptance rate	Restricted amount of the aircraft entering into the terminal area, crossing at terminal gates and meter fixes, landing on the runway, etc.	Center's host computer	4DT-DP is initialized/acceptance rate is updated
7	Blocked interval	Period of time restricting any aircraft from crossing specific meter fixes and/or landing on the runway	Center's host computer	4DT-DP is initialized/blocked interval is updated
No. 8–9 sequence constraints				
8	PBO information	Types of PBO (e.g., ASAS and RNP) and status of the operation (e.g., executed or not)	Center's host computer	4DT-DP is initialized/PBO information is updated
9	Air traffic controller's intention		GUI	Air traffic controller changes the arrival schedule

Table 2 Inputs to 4DT-TS

No	Input information	Information details	Sources	Update timing
1	Flight objects	Aircraft ID, types, characteristics, flight plans, updated 4DT, etc.	Center's host computer	4DT-TS is initialized/4DT is updated
2	4DT route	Analyzed route from the aircraft current position to the runway following the assigned 4DT	4DT-RA	All updates of 4DT-RA
3	Track data	Aircraft tracks in radar data	Center's host computer	All updates of 4DT-TS
4	Weather information	Current and predicted weather information including wind prediction	Center's host computer	All updates of 4DT-TS
5	Aircraft performance data	Aircraft weight, setting of the cost indices, time constraints at specific points, speed restrictions, and any other performance matrixes which determine descent/climb performance (e.g., values of climb thrust, speed brake, flap angle, and anti-ice settings)	Center's host computer	All updates of the aircraft performance data
6	ETA on board	FMS generated ETA in-flight	Airborne FMS	When the ETA is downlinked

resembles the given sequence constraints by ATCos. The intention of the ATCos should be prioritized in the arrival sequencing.

Table 2 summarizes the five types of information inputs which should be shared with 4DT-TS for ETA estimation. "Flight objects," "4DT route," "track data," and "weather information" are required to improve the accuracy of the ETA estimation. "Aircraft performance data" provides the constraints of the ETA estimation. This includes aircraft weight, cost indices, time constraints, and other settings on the climb and descent phases since they affect the flight altitude and airspeed, which determine the accuracy of the ETA estimation. "ETA on board" is the ETA downlinked from the airborne-equipped FMS via datalink. The on-board FMS estimates the ETA at assigned waypoints and the runway threshold. It helps to compensate estimation errors between ground-based ETA estimated in 4DT-TS in case it has a periodic downlink to the ground.

Table 3 acquires three information inputs required for updating *4DT analyzed routes* in 4DT-RA. Since 4DT-RA updates the *4DT analyzed routes*, "flight objects," "track updates," and "aircraft performance data" should be updated to the 4DT-RA.

Table 3 Inputs to 4DT-RA

No	Input information	Information details	Sources	Update timing
1	Flight objects	Aircraft ID, types, characteristics, flight plans, updated 4DT, etc.	Center's host computer	4DT-RA is initialized/4DT is updated
2	Track updates	Aircraft tracks in radar data	Center's host computer	All updates of 4DT-RA
3	Aircraft performance data	Aircraft weight, setting of the cost indices, time constraints at specific points, speed restrictions, and any other performance matrixes which determine descent/climb performance (e.g., values of climb thrust, speed brake, flap angle, and anti-ice settings)	Center's host computer	All updates of the aircraft performance data

When the 4DT-DP allocates the arrival runway, the updated runway information should be stored in the flight objects.

4.3 Air-Ground Harmonization

The future ATM system is required to handle the increasing number of airborne-based operations and conventional ground-based operations under a mixed equipage scenario. One issue to be clarified in achieving the harmonization of ground- and airborne-based operations is how to maximize the potentials of airborne-based operations by centralizing the mixed equipage air traffic on the ground by utilizing automation supports, as represented by E-AMAN.

The implementation of 4D TBO requires CDM between air to ground and ground to ground. The 4DT should be shared between FMS in the air and E-AMAN on the ground. The ideal operational method would be the use of datalink communication for air-ground harmonization.

Minimizing estimation errors between estimated and actual wind improves the accuracy of ETA estimation in 4DT-TS. One suggestion is to compensate the ground-estimated ETA by downlinking ETA from the FMS [13]. It is required to determine the frequencies of the updates in order to satisfy the accuracy of the ETA estimation.

Various implementations of the airborne-based operations will be carried out by the year 2030 using ASAS equipment, high-level RNP, and required time of arrival (RTA) functions in the FMS. These airborne-based operations execute automatic speed control for the purpose of maintaining aircraft separation and minimizing delay in the arrival time. The authors have been analyzing the feasibility and benefits of the airborne-based operations focusing on RTA and FIM operations [6–8]. The latest RTA function is designed to control the airspeed in the descent phase where its predecessor lacked the capability. FIM speed control is capable of achieving precise time spacing between its successive aircraft even on the descent path. The airborne-based speed control has the potential to minimize arrival delay. The proposed E-AMAN needs to share the status of the airborne-based operations, for instance, what aircraft possesses which airborne equipage and executes what types of operation, in order to generate BEBS scheduling in the mixed equipage air traffic.

In this manner, the E-AMAN is expected to centralize conventional ground-based and airborne-based operations.

4.4 Human-Computer Interaction

A highly automated system is expected to collaborate with human operators via human-computer interaction (HCI). The proposed E-AMAN is a ground automation system which alternates a part of intelligent tasks of ATCo in arrival operations. ATCos and pilots are expected to achieve arrival sequences and STA generated by the E-AMAN without any *automation surprise*. Thus, the HCI design should take its operational usability into consideration.

Conventionally, the HCI has been studied and developed to support ATCos to follow the scheduling procedures suggested by ground advisory tools. Figure 9 shows one of the examples of the TMA timelines displayed to ATCos. This timeline compares STA and ETA at assigned points and shows the amount of delays in the arrival time. Air Traffic Technology Demonstration-1 (ATD-1) project [14] was shown on the display design targeting the harmonization of ground-based and

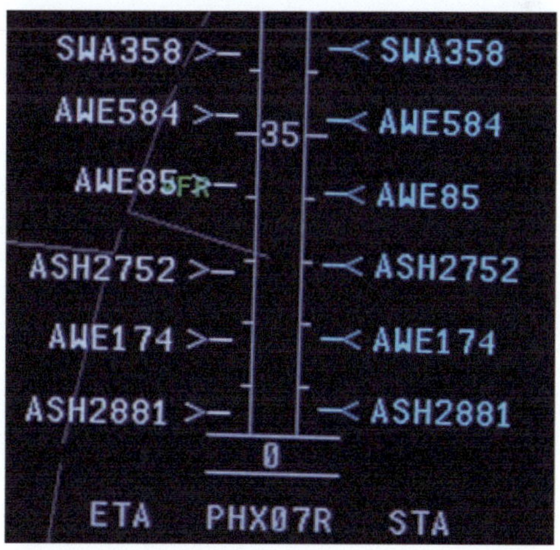

Fig. 9 Comparison of STA and ETA on a radar display [14]

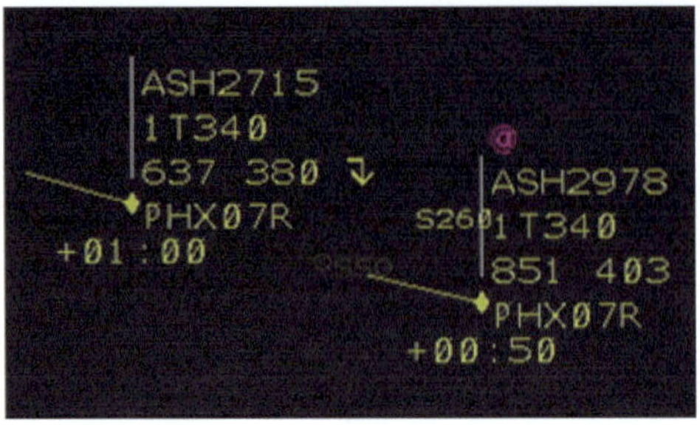

Fig. 10 FIM aircraft on a radar display [14]: "@" is given to the FIM aircraft

Fig. 11 EFB (electronic flight bag) [14]

airborne-based operations including FIM and RNP operations. Figure 10 illustrates one of the examples displayed on the radar display. As shown in Fig. 10, a specific symbol is shown above the FIM aircraft tag on the radar display. FIM status in operation (e.g., executed or terminated) is indicated to ATCos on the ground by displaying these symbols in different colors over the aircraft tag. In the cockpit of the FIM aircraft, an electronic flight bag (EFB) is equipped on the window side of the pilot's seat as shown in Fig. 11. The pilot inputs the required information to the FIM execution through the EFB. The EFB links to the FMS and shares the flight information. Target speed command for the FIM operation is shown to the pilot via the EFB. User-friendly design of the human-computer interaction is required in the above described scenarios.

5 Concluding Remarks

This paper presented a future vision of TBO-based arrival management and technologies to support the E-AMAN implementation. Operational concepts of future E-AMAN collaborating with 4D TBO were proposed, targeting the year 2030. Technical challenges were discussed to implement the E-AMAN in the operations. Initially, concepts and past works on 4D TBO and E-AMAN were reviewed. Then, operational concepts were proposed and expected benefits were discussed. Information sharing, air-ground harmonization, and user-friendly design of human-computer interaction were chosen as key technologies to design the future E-AMAN.

The growth of the Asian economy is assumed to double the global air traffic by the year 2030. This study enhances the future E-AMAN design and develops an efficient E-AMAN targeting the expected amount of arrivals to the Tokyo metropolitan area in 2030. The significance of the proposed future E-AMAN will be evaluated by both fast-time simulations and human-in-the-loop simulation experiments.

Acknowledgments This research is conducted as a part of research programs for Promoting Technical Development of Transportation funded by the Ministry of Land, Infrastructure, Transport and Tourism in Japan.

References

1. International Civil Aviation Organization (ICAO), "Global Air Traffic Management Operational Concepts", ICAO Doc 9854, AN/458, 2005.
2. M. Brown, M. Fujita, Y. Fukuda, S. Inoue, H. Hirabayashi, and S. Nagaoka, "Full 4D Trajectory Based Operations Concept Study", APISAT 2013, Nov. 2013.
3. S. Nagaoka and M. Brown, "A Difficulty Index for Air Traffic Control Based on Potential Conflicts", 27th European Conference on Operational Research, July 2015.
4. N.K. Wickramasinghe, Y. Miyazawa, A. Harada, T. Kozuka, and S. Shigetomi, "Flight Trajectory Optimization for Operational Performance Analysis of Jet Passenger Aircraft", APISAT 2013, Nov. 2013.
5. A. Harada, T. Kozuka, Y. Miyazawa, N.K. Wichramasinghe, M. Brown, and Y. Fukuda, "Analysis of Air Traffic Efficiency using Dynamic Programming Trajectory Optimization", 29th ICAS, Sep. 2014.
6. E. Itoh, K. Uejima, N.K. Wickramasinghe, H. Hirabayashi, and S. Fukushima, "Continuous Descent Operation Following Fixed-flight Path Angle from Oceanic Air Route to Tokyo International Airport", AIAA Science and Technology Forum and Exposition 2016, Jan. 2016.
7. E. Itoh and K. Uejima, "Applying Flight-deck Interval Management Based Continuous Descent Operation for Arrival Air Traffic to Tokyo International Airport", ATM seminar 2013, June 2013.
8. E. Itoh, K. Uejima, K. Kakichi, and S. Suzuki, "Modeling and Simulation Study on Airborne-based Energy Saving Arrivals to Tokyo International Airport", AIAA Guidance, Navigation, and Control Conference, AIAA 2013-4779, Aug. 2013.

9. International Civil Aviation Organization (ICAO), "Continuous Descent Operations (CDO) Manuals", ICAO Doc 9931, AN/476, 2010.
10. F. Neuman, and H. Erzberger, "Analysis of Delay Reducing and Fuel Saving Sequencing and Spacing Algorithms for Arrival Traffic", NASA/TM103880, Oct. 1991.
11. G.L. Wong, "The Dynamic Planner: The Sequencer, Scheduler, and Runway Allocator for Air Traffic Control Automation", NASA/TM-2000-209586, April 2000.
12. H. Erzberger and E. Itoh, "Design Principles and Algorithms for Air Traffic Arrival Scheduling", NASA/TP-2014-218302, May 2014.
13. Y. Fukuda, M. Shirakawa, A. Senoguchi, "Development and Evaluation of Trajectory Prediction Model", 27th International Congress of the Aeronautical Sciences, ICAS 2010–5.6.1, Sept. 2010.
14. NASA Ames Research Center AF Division, "ATD-1 HITL simulation", https://www.youtube.com/watch?v=ngKazVQN4BI, March 2015.

Part III
Optimization of Air Traffic

Quantitative Analysis of Conflict Between Aircraft by Using Radar Track Data

Tomoyuki Kozuka and Yoshikazu Miyazawa

Abstract As the demand for air traffic increases, new air traffic management systems are needed. In the system, it is expressly necessary to avoid conflict between aircraft. Efficient methods to prevent conflict are currently based on the expertise of air traffic controllers. Hence, a two-step analysis method is described in this study. First, it is shown using a proposed method with en route radar track data that no conflicts occurred in the airspace over Japan during the period in question. Second, instructions provided to pilots by the air traffic controllers to solve conflicts are estimated from the same radar data, and the results allow some groups to be rearranged in terms of avoidance procedures. These results will be useful references when developing an automated system of conflict detection and resolution in the future.

Keywords Conflict • Radar track data • Flight trajectory • Separation index

1 Introduction

Global air traffic is increasing with significant air demand in various parts of the world, and this trend is predicted to continue over the next few decades [1]. In Japan, the demand in 2027 is anticipated rising to 1.5 times that in 2005 in a report on Collaborative Actions for Renovation of Air Traffic Systems (CARATS) [2]. If air traffic grows at the present pace, new issues will arise. The number of air traffic controllers in Japan has declined slightly within the past 10 years, but the workload of air traffic controllers has increased further. It is expressly important to avoid conflict between aircraft, and much research has been conducted on this subject. For example, references [3, 4] suggest conflict resolution using actual data from air traffic control.

Furthermore, there has been much research on operating airliners to cope with high economic efficiency while ensuring safety of their users and customers [5–8].

T. Kozuka (✉) • Y. Miyazawa
Graduate School of Engineering, Kyushu University, Fukuoka, Japan
e-mail: moonofringbear75@gmail.com; miyazawa@aero.kyushu-u.ac.jp

Electronic Navigation Research Institute, *Air Traffic Management and Systems II*,
Lecture Notes in Electrical Engineering 420, DOI 10.1007/978-4-431-56423-2_8

In contrast, papers [9–10] on the subject of conflict resolution in Japan are less numerous than those on this subject in Western countries. We are motivated by this situation and aim for research on conflict-free flight trajectories.

To solve the problem of conflict between aircraft, we should refer to the expertise air traffic controllers. In this study, there are two steps to find the actual procedures of air traffic controllers for conflict resolution. First, it is shown using radar track data that no conflicts occurred in Japanese airspace during the analyzed term. Second, we estimate the instructions from air traffic controllers to pilots. The results are to be used to progress research on conflict resolution in further studies.

2 Estimation of Conflict

2.1 CARATS Open Data

"CARATS Open Data" are used to predict conflicts between aircraft and to estimate air traffic controllers' instructions for resolution. The data are from radar tracks obtained by the Japan Civil Aviation Bureau (JCAB) and include the time histories of commercial flights' position data within the Fukuoka Flight Information Region (FIR), as shown in Fig. 1. Red dashed lines indicate the boundaries of the four Area Control Centers (ACCs). An outline of the data is listed in Table 1. The data

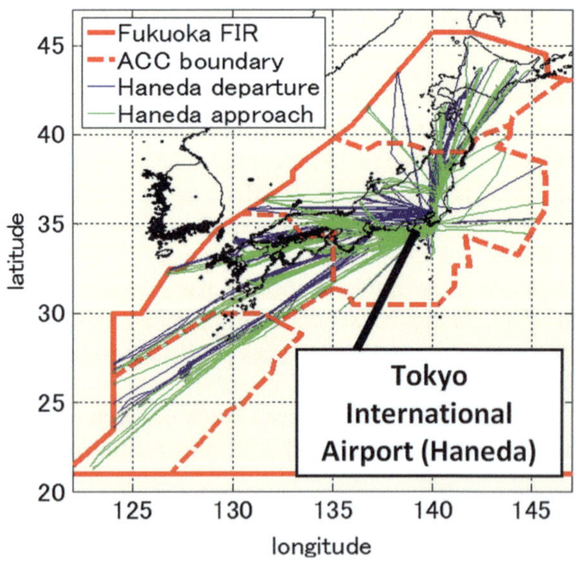

Fig. 1 Radar tracks from CARATS open data

Table 1 CARATS open data

Term	May 7, 2012–May 13, 2012
	July 9, 2012–July 15, 2012
	September 3, 2012–September 9, 2012
	November 5, 2012–November11, 2012
	January 7, 2013–January 13, 2013
	March 4, 2013–March 10, 2013 (total 42 days)
Target	Airliners (only IFR)
Count	About 3600 [flights/day]
Parameters	Time, hypothetical flight number, latitude, longitude, aircraft type
Data cycle	About 10 [s]

originate from a total of 42 days from May 7, 2012, to March 10, 2013 (one week per odd-numbered month).

2.2 *Preprocessing the Data*

In general, radar track data inevitably contain various errors, and CARATS Open Data include some errors and unexpected values. One reference analyzes the accuracy of CARATS Open Data and shows their errors [11].

Our preprocessing consists of four steps to reduce the number of errors and unexpected values. The first step is to get rid of all values that have position error. The three-axis acceleration of the aircraft from three sets of position data is calculated. When the acceleration exceeds a threshold, the third position dataset is removed. The second step is prediction of accurate time. The time included after the decimal point is predicted using a linear search method in terms of acceleration. The third step is to eliminate unexpected values. If the unexpected values of altitude data are found and the data have a vertical rate exceeding the threshold, then they are removed. The fourth step is hand-control processing. When conflicts are identified with a method mentioned in the next subsection, the data are taken away only if the reason for the conflict is the unexpected value.

2.3 *The Method of Conflict Estimation*

To prepare our analysis, CARATS Open Data are interpolated to estimate conflict every 10 s. Combinations of two aircraft are made 10 s. From these combinations, the distance between two aircraft is calculated using Eqs. (1) and (2) from the position data. The earth can be approximately represented as a sphere. R [m], η, ϕ [deg], and θ [deg] are, respectively, the earth radius, the angle between two positions, latitude, and longitude. Subscripts 1 and 2 express two aircraft. | | denotes norm of the vector.

$$d = R\eta = 2R\sin^{-1}\frac{1}{2}\left|\begin{pmatrix}\cos\phi_2\cos\theta_2\\ \cos\phi_2\sin\theta_2\\ \sin\phi_2\end{pmatrix} - \begin{pmatrix}\cos\phi_1\cos\theta_1\\ \cos\phi_1\sin\theta_1\\ \sin\phi_1\end{pmatrix}\right| \tag{1}$$

$$\Delta h = h_2 - h_1 \tag{2}$$

To judge whether a conflict between aircraft is occurring, a safety zone is defined, which is shown in Fig. 2. In this study, $d_0 = 5$[NM] and $h_0 = 1\,,000$[ft] are assigned. We determine the size of the safety zone from Aeronautical Information Publication Japan (AIP JAPAN) and the Japanese rules of air traffic control because our range of analysis is mainly the control area (above 12,000 [ft]). Both materials are published by JCAB.

The separation index shown in Eq. (3) is defined to quantify the conflict using the previous distance. Relative distances are normalized by d_0 and h_0 to gain the separation index:

$$I_{\mathrm{spr}} = \max\left\{\frac{|\Delta h|}{h_0}, \frac{d}{d_0}\right\} - 1. \tag{3}$$

Ellipsoid zone is used to quantify the distance between two aircraft in other references [12], but this circular cylinder safety zone is more consistent with the present rule of aircraft separation. I_{spr} is the separation index, which indicates the relative position between two aircraft as follows:

$$\begin{array}{ll}(1)\ I_{\mathrm{spr}} > 0: & \text{outside.}\\ (2)\ I_{\mathrm{spr}} = 0: & \text{on surface.}\\ (3)\ I_{\mathrm{spr}} < 0: & \text{inside.}\end{array} \tag{4}$$

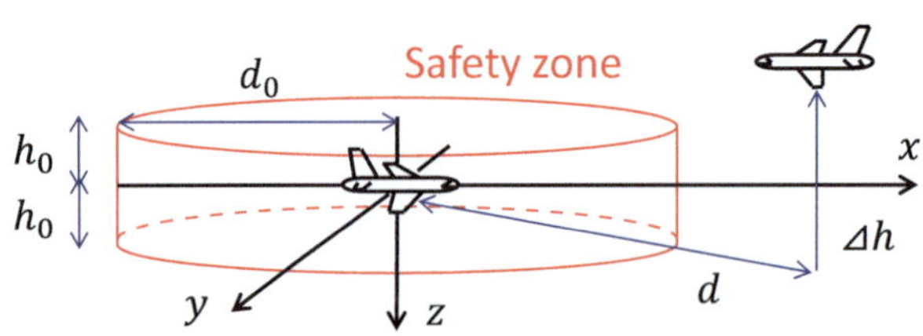

Fig. 2 Safety zone

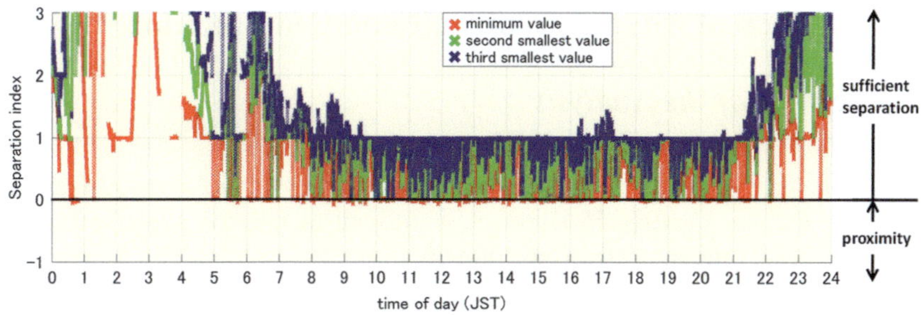

Fig. 3 Conflict estimation (January 8, 2013)

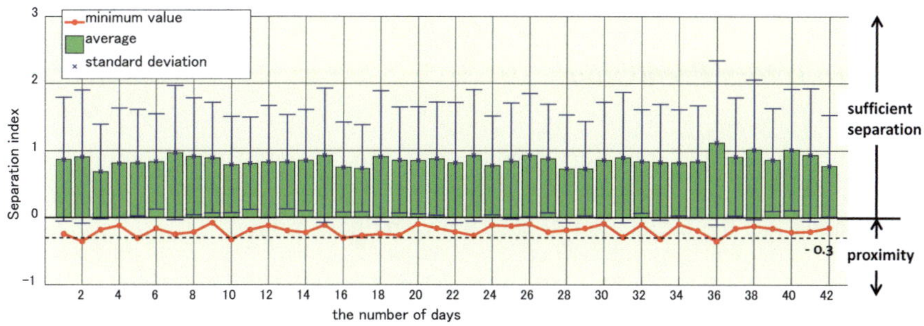

Fig. 4 Conflict estimation (42 days, May 7, 2012–March 10, 2013)

Red points in Fig. 3 show the minimum value of the separation index every 10 s, green points are the second-smallest, and blue points are third. Red points sometimes exist in the negative domain, but the data include a few errors. Because of error value or overshoot, we think it is a critical conflict when separation index is below −0.3 by far. Thus, Fig. 3 indicates that no conflict occurs in Fukuoka FIR every 10 s.

Figure 4 shows the results of conflict estimation in 42 days from May 2012 to March 2013. A red solid line indicates the minimum value of the separation index each day. Green bars are the averages of the minimum values taken each 10 s of a day. Blue bars are the standard deviation of the minimum value each day. The black dashed line is the value −0.3 of the separation index. If the value is above −0.3, it is practically considered not a critical conflict even if below 0. The results indicate two things: first, no conflict took place in Fukuoka FIR during the analyzed days. Second, the cases of the minimum values are not generated frequently because the minimum values are outside of standard deviation.

3 Estimation of Aircraft Conflict Resolution Procedure

3.1 Purpose of the Analysis

The goal is to devise optimal flight trajectories in terms of fuel consumption and flight time without conflict. However, it is difficult to develop a method that is suitable for actual situations to avoid conflict.

We extract the instructions of actual air traffic controllers pertaining to separation between aircraft from the data. The controllers' expertise will be referred to in studies on creating conflict-free flight trajectories.

3.2 Trajectory Prediction

In this method, simple trajectory prediction is used to find future conflict. First, ground speed and true tracks are calculated from position data by Eqs. (5) and (6). It is assumed that the target aircraft have uniform linear motion:

$$\tan(\psi_t) = \frac{\cos\phi_t \sin(\theta_t - \theta_{t-\Delta t})}{\cos\phi_{t-\Delta t}\sin\phi_t - \sin\phi_{t-\Delta t}\cos\phi_t\cos(\theta_t - \theta_{t-\Delta t})}; \tag{5}$$

$$\phi_{\text{prd}} = V_{GS}t_{\text{prd}}\cos(\psi_t)/a + \phi_t, \quad \theta_{\text{prd}} = V_{GS}t_{\text{prd}}\sin(\psi_t)/b + \theta_t; \tag{6}$$

$$h_{\text{prd}} = h + \frac{dh}{dt}t_{\text{prd}}. \tag{7}$$

Here, ψ_t, ϕ, θ, ϕ_{prd}, θ_{prd}, V_{GS}, h_{prd}, h, t_{prd}, a, and b are true track [deg], latitude [deg], longitude [deg], predicted latitude [deg], predicted longitude [deg], ground speed [m/s], predicted altitude [ft], altitude [ft], predictive period [s], and transformation coefficients which translate distances to latitude angle and longitude angle, respectively. a and b are constant numbers from $a = R$ and $b = R\cos(\phi_J)$. ϕ_J is representative latitude around Tokyo. Subscript t means the present time, and $t - \Delta t$ means one step earlier time of data sampling. The predicted separation index is obtained from predicted longitude and latitude by using Eqs. (1) and (2). The predictive period is 180 [s]. To decide the predictive period, we refer to a conflict-alerting system of the Radar Data Processing (RDP) system. In Fig. 5, the present positions and predicted trajectories are plotted.

3.3 Algorithm of the Method

Actual aircraft conflict resolution procedures are estimated from CARATS Open Data, with the separation index defined in Sect. 2.3. First, the separation index at the

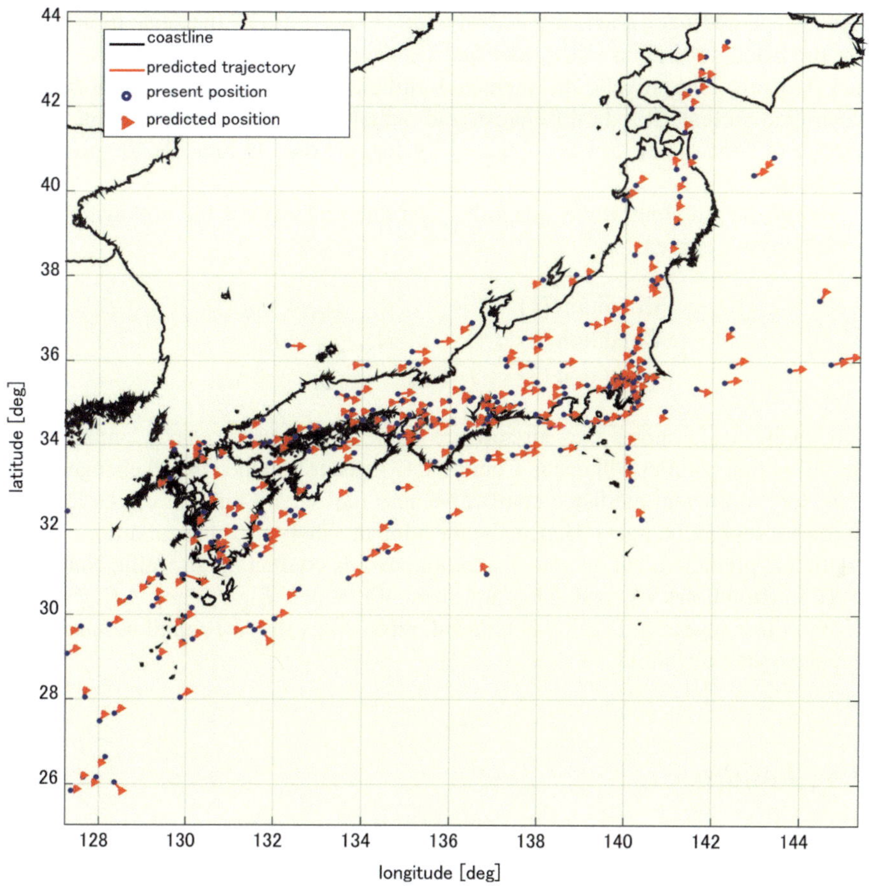

Fig. 5 Predicted positions in Fukuoka FIR

present time is calculated. Second, the following three values are computed using Eqs. (8), (9), and (10):

$$\dot{I}_{\mathrm{spr}} = \frac{dI_{\mathrm{spr}}}{dt}. \tag{8}$$

$$I_{\mathrm{spr\_prd}} = \max\left\{\frac{|\Delta h_{\mathrm{prd}}|}{h_0}, \frac{d_{\mathrm{prd}}}{d_0}\right\} - 1. \tag{9}$$

$$I_{\mathrm{spr\_real}} = \max\left\{\frac{|\Delta h_{\mathrm{real}}|}{h_0}, \frac{d_{\mathrm{real}}}{d_0}\right\} - 1. \tag{10}$$

$\dot{I}_{spr}$, I_{spr_prd}, and I_{spr_real} are, respectively, the change ratio of the separation index, the predicted separation index, and the real separation index. Δh_{prd}, d_{prd}, Δh_{real}, and d_{real} are, respectively, the predicted altitude difference, predicted horizontal separation, actual altitude difference, and actual horizontal separation, all at the predictive time. They can be obtained by using Eqs. (1) and (2) for predicted positions.

Using the previous three quantities, aircraft that satisfy the following three conditions are chosen:

$$\begin{array}{ll} \text{Condition (1)} & \dot{I}_{spr} < 0. \\ \text{Condition (2)} & I_{spr\_prd} < I_{jdg}. \\ \text{Condition (3)} & I_{spr\_real} - I_{spr\_prd} > \Delta I_{jdg}. \end{array} \tag{11}$$

Condition (1) shows two aircraft are approaching each other. Condition (2) describes that conflict will occur. Condition (3) means that the aircraft changed own trajectory to prevent predicted conflict because real separation index is larger than predicted separation index. Hence, the meaning of satisfying three conditions is that conflict is predicted, but, in actual situation, avoids conflict by air traffic controller (ATC) instructions. I_{jdg} and ΔI_{jdg} are threshold values. $I_{jdg} = 0$ and $\Delta I_{jdg} = 1$ are given in this research. If ΔI_{jdg} is reduced, more cases may be found because they depend on the threshold.

3.4 Results

Estimations of aircraft conflict resolution procedures are classified into three patterns. The patterns are indicated in Table 2. However, values of velocity have their own oscillation because CARATS Open Data include error. As it is difficult to find changing velocity for conflict resolution, pattern (3) is excluded in this study.

Table 2 Patterns of resolution procedures (00:00 to 24:00 on January 7, 2013)

Patterns	Situations	Number
(1)	Changing altitude	119
(2)	Changing heading	73
(3)	Changing velocity	–
(4)	Others	22

Total extracting cases are 185 from 00:00 to 24:00 in single day, January 7, 2013. "Number" permits overlap when the case consists of two ATC instructions

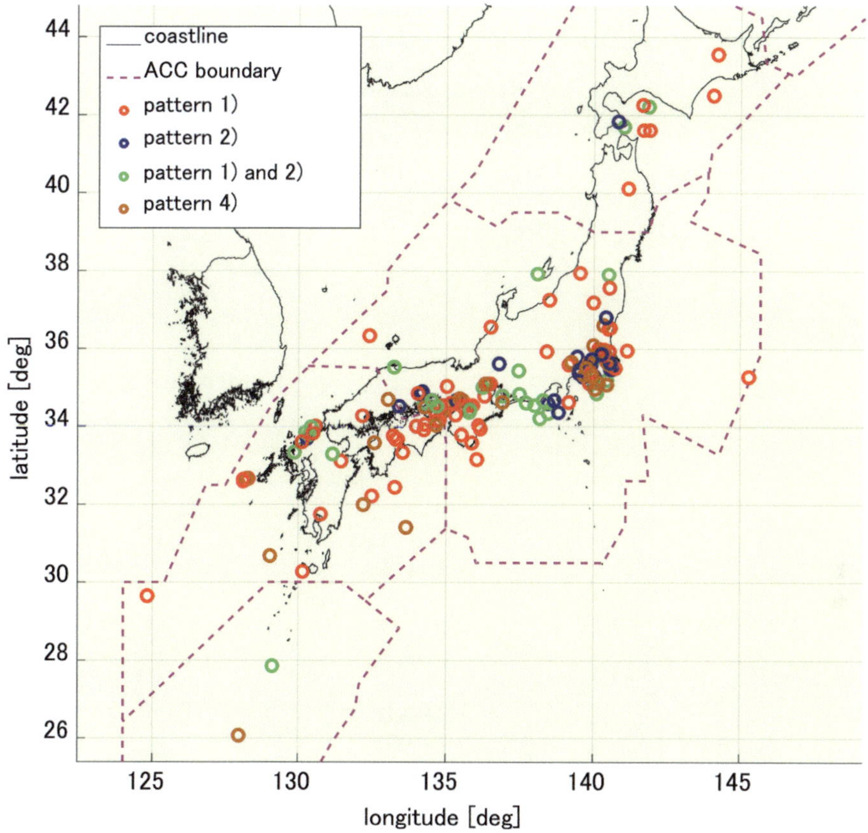

Fig. 6 Positions predicting conflict (00:00 to 24:00 on January 7, 2013)

Results of estimating procedures are shown in Figs. 6, 7, 8, 9, and 10 and Table 2. The range of analysis is from 00:00 to 24:00 on January 7, 2013. The item "Number" in Table 2 indicates the number of predicted ATC instructions. "Number" permits overlap in Table 2 when the case consists of two ATC instructions. There are 185 cases in the range of analysis. The results indicate that pattern (1) is chosen the most frequently among the three patterns. This tendency is reasonable because ensuring altitude is clear and easy for controllers and pilots. In Fig. 6, positions of conflict which we estimated are shown. In Figs. 7, 8, 9, and 10, the cases of patterns (1) and (2) are shown. Figures 7 and 8 are at the same case, and both aircraft leveled off to ensure the altitude difference by altitude change instruction. Similarly, Figs. 9 and 10 are also at the same case, and cruise aircraft changed own heading to horizontal distance by heading change instruction. Although conflict is predicted, two aircraft can avoid conflict by following ATC instructions.

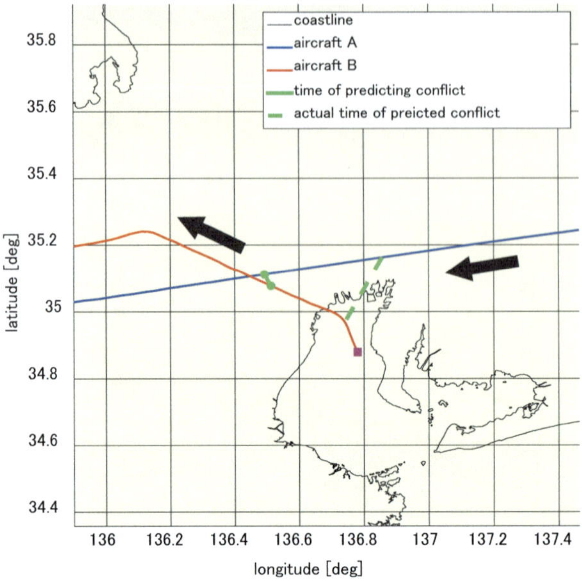

Fig. 7 Flight track of pattern (1)

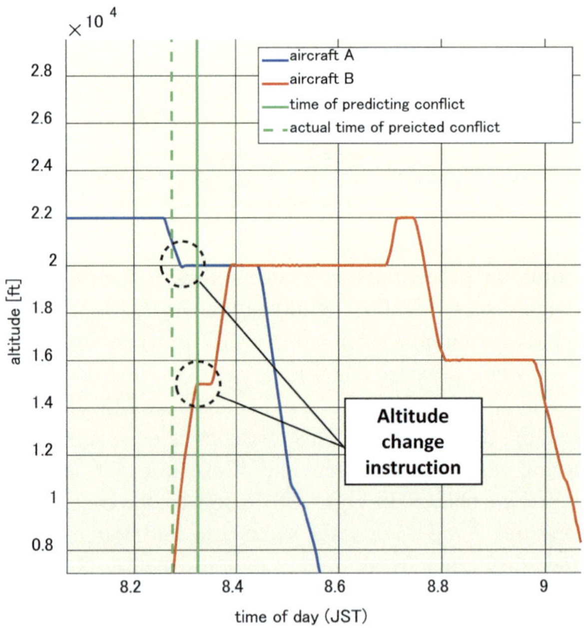

Fig. 8 Time history of altitude of pattern (1)

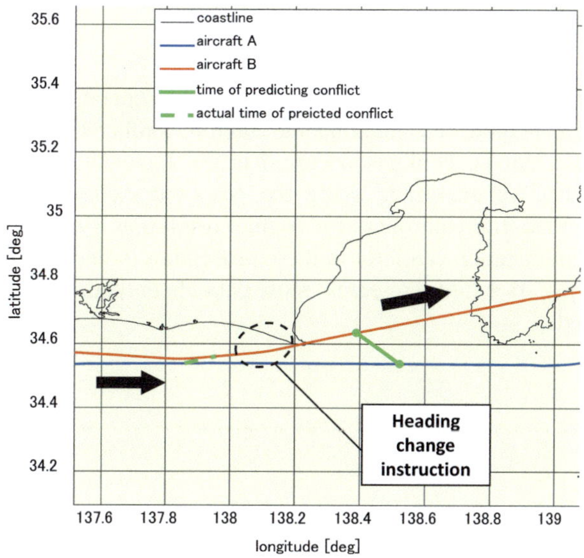

Fig. 9 Flight track of pattern (2)

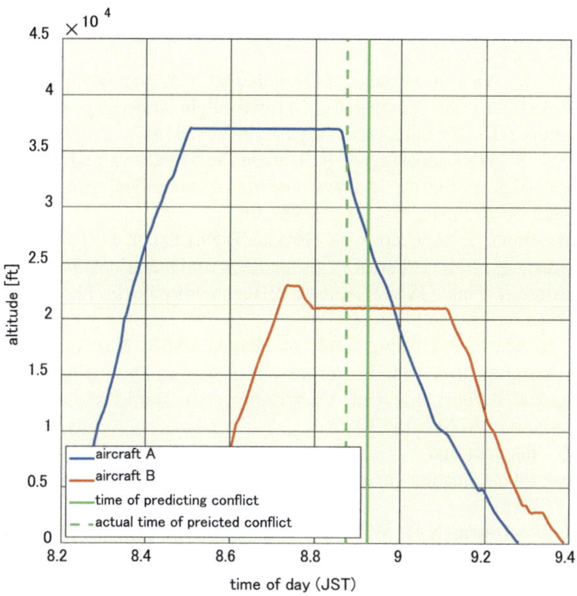

Fig. 10 Time history of altitude of pattern (2)

4 Conclusion

We defined the separation index as the amount of separation between aircraft and proposed a method of estimating the aircraft conflict resolution procedure. Results indicate that no conflicts occurred in the Fukuoka FIR region through the 42 days that we analyzed. Using the proposed method, we obtained the procedures of air traffic controllers for conflict resolution. Hence, future research on trajectory optimization can refer to these procedures in developing conflict-free flight trajectories. As a future research, more data of conflict resolution procedures will be gathered.

References

1. The Boeing Company: Current Market Outlook 2013-2032. https://www.iaa.ie/docs/default-source/misc/boeing_current_market_outlook_20131.pdf. Accessed 06 September 2016
2. Study Group for the Future Air Traffic Systems: Long Term Vision for the Future Air Traffic Systems, CARATS Collaborative Actions for renovation of Air Traffic Systems. http://www.mlit.go.jp/common/000128185.pdf. Accessed 06 September 2016
3. Paielli, R. A.: Tactical Conflict Resolution Using Vertical Maneuvers En Route Airspace. J. Aircraft, 45, 6 (2008), p.2111-2119.
4. Paielli, R. A., “Evaluation of Tactical Conflict Resolution Algorithms for Enroute Airspace”, J. Aircraft, 48, 1, 324–330 (2011)
5. Y. Miyazawa, N. K. Wickramasinghe, A. Harada and Y. Miyamoto: Dynamic Programming Application to Airliner Four Dimensional Optimal Flight Trajectory. AIAA Guidance, Navigation, and Control (GNC) Conference, August 19–22 (2013)
6. Y. Miyamoto, N. K. Wickramasinghe, A. Harada, Y. Miyazawa and K. Funabiki: Analysis of Fuel-Efficient Airliner Flight via Dynamic Programming Trajectory Optimization. Trans. JSASS Aerospace Tech. Japan, Vol. 11, 93–98 (2013)
7. N.K. Wickramasinghe, Y. Miyamoto, A. Harada, T. Kozuka, S. Shigetomi, Y. Miyazawa, M. Brown and Y. Fukuda: Flight Trajectory Optimization for Operational Performance Analysis of Jet Passenger Aircraft. Trans. JSASS Aerospace Technology Japan, 12, APISAT2013, a17–a25 (2014)
8. Y. Miyazawa, H. Matsuda, S. Shigetomi, A. Harada and T. Kozuka: Potential Benefits of Arrival Time Assignment, Dynamic Programming Trajectory Optimization applied to the Tokyo International Airport. Eleventh USA/Europe Air Traffic Management Research and Development Seminar, ATM2015 (2015)
9. K. Fukuoka, N. Takeichi and Y. Nakamura: Conflict Avoidance Using Flight Speed Control in a High Density Air Corridor. (in Japanese), Trans. JSASS Aerospace Tech. Japan, Vol. 61, No.5, 119–124 (2013)
10. T. Kozuka, Y. Miyazawa, N.K. Wickramasinghe, M. Brown and Y. Fukuda: A Study on Estimation of Aircraft Interference Resolution Procedure by Surveillance-Derived Flight Trajectory. (in Japanese), Trans. JSASS Aerospace Tech. Japan, Vol. 14, 77–83 (2015)
11. S. Shigetomi, T. Kozuka, Y. Higuchi, K. Yoda and Y. Miyazawa: Evaluation of CARATS Open Data Accuracy. (in Japanese), Trans. JSASS Aerospace Tech. Japan, Vol 15, 15–21 (2016)
12. EUROCONTROL: EUROCONTROL Guidance Material for Short Term Conflict Alert Appendix A: Reference STCA System. https://www.eurocontrol.int/sites/default/files/field_tabs/content/documents/nm/safety/appendix-a-reference-stca-system.pdf.

Optimizing the Design of a Route in Terminal Maneuvering Area Using Branch and Bound

Jun Zhou, Sonia Cafieri, Daniel Delahaye, and Mohammed Sbihi

Abstract The sharp increase in air traffic flow causes traffic congestion in airspaces near airports, called Terminal Maneuvering Areas (TMA). The departure and arrival traffic of airports follow predesigned routes named standard instrument departure (SID) routes and standard terminal arrival routes (STAR). Optimizing these routes is crucial to regulate air traffic. Currently, SIDs and STARs are designed manually, based on the airport layout and nearby constraints. The objective of this research is to propose a methodology for designing an arrival/departure route in TMA, taking into account some constraints including obstacle avoidance. The shape of a route in horizontal plan is a succession of arcs of circles and segments. The originality of our study is, on the one hand, that the horizontal route is associated with a cone in vertical plan enveloping all ascent (or descent) aircraft profiles, and on the other hand, a branching strategy in a branch and bound (B&B) framework tailored on the problem is proposed.

Keywords TMA • SID/STAR design • Obstacle avoidance • Modeling • Branch and Bound

1 Introduction

The continuous increase in air traffic flow density causes traffic congestion in the areas surrounding airports, thus affecting the normal operation of air traffic. An area surrounding one or more neighboring airports is called Terminal Maneuvering Area (TMA), and it is designed to handle aircraft arriving to and departing from the airports. Optimizing departure and arrival procedures in TMA is therefore crucial to regulate air traffic flows. Most of the airports have predesigned procedures indicating how aircraft depart from or arrive to airports. These procedures are called standard instrument departure (SID) routes and standard terminal arrival route (STAR). A SID is a flight route followed by aircraft after takeoff from an

J. Zhou (✉) • S. Cafieri • D. Delahaye • M. Sbihi
ENAC, Toulouse, France
e-mail: junzhou@recherche.enac.fr; sonia.cafieri@enac.fr; daniel.delahaye@enac.fr; mohammed.sbihi@enac.fr

Electronic Navigation Research Institute, *Air Traffic Management and Systems II*, Lecture Notes in Electrical Engineering 420, DOI 10.1007/978-4-431-56423-2_9

airport until the start of en route phase. A STAR is a route which connects the last en route way point to the Initial Approach Fix. Currently, SID/STARs are designed manually according to operational requirements (ICAO Doc 8168), taking into account airport layout and nearby constraints. However, this kind of design is generally not very efficient and not expected to optimize any specific criterion. The objective of this work is to automatically design SID/STARs in 3D with respect to certain optimization criteria. Being this study at a strategic level, only static obstacles are taken into account. Rather than considering an individual flight, we deal with flows of flights, that is to say the flights following the same SID (or STAR) belong to the same flow.

The considered problem is in the framework of path planning. Specifically, it is a route design problem: as contrarily to trajectory design, we aim at designing routes that are not associated with the notion of time. The problem of path planning has been studied since the 1980s in the robotic domain [3, 9]. Nowadays planning optimal aircraft paths becomes a rich and dynamic research area; some approaches have been summarized in [1]. In particular, computing the shortest path between two points, given a number of obstacles, is one of the most extensively studied topics [4, 5, 7, 8, 10]. Most of these works search for the shortest path in a horizontal plan, while in our study we consider searching for the shortest path in 3D. Moreover, in several works the obstacle is modeled as polygon [5, 8, 10] and in a few works as circle [4, 7]. In the present study, we model the obstacle as cylinder in 3D, and the projection to the horizontal plan is in the form of circle.

Specifically in aircraft path designing domain, even though there is a large number of researches, route design in TMA is a particular problem for which to our knowledge there is not a rich literature. In [6] the author designs terminal routes getting around obstacles with a modified A* algorithm. In our study, we design a route not only getting around obstacles but also allowing level flights. Indeed, imposing a level flight in vertical plan is also an effective way to avoid obstacles, as it enriches the possible maneuvers and corresponds to what is done in the reality in a TMA. The different ways to avoid obstacles allow us to define specific branching strategies in a branch and bound tailored on the problem, that is, another contribution of this work.

This paper is organized as follows. Section 2 introduces the route and obstacle modeling. Section 3 presents the proposed approach to solve the problem. Section 4 gives some preliminary simulation results. Finally, Sect. 5 draws conclusions and proposes future directions.

2 Problem Modeling

TMA is one of the most complex types of airspace. Many constraints have to be satisfied, falling into two categories: operational constraints related to air traffic operations (such as obstacle avoidance and flyable routes) and environmental constraints (such as noise abatement). SID/STARs are designed to satisfy these constraints and to deal with the dense traffic converging to and diverging from airports. The constraints in TMA make the SID/STARs design a very complex problem. Therefore, in this study we consider the simpler subproblem of designing a single route avoiding obstacles and satisfying some other operational constraints. The obstacles in TMA could be mountains, cities, military area, etc. In the following, we present the way we model routes and obstacles.

A 3D route γ is defined by two elements: a curve γ_H in a horizontal plan which is composed by a succession of arcs of circles (to bypass obstacles) and segments (to connect tangentially two arcs) and a cone γ_V in a vertical plan that contains all ascent (or descent) profiles of the aircraft flying on this route. The cone is defined by two straight lines whose slopes are the minimum and maximum values of the takeoff (or landing) rate of the aircraft on this route. The idea of taking a cone that contains all vertical profiles is inspired by the behavior illustrated in Fig. 1, which shows some real takeoff data in runway 08L of Paris CDG airport. From the figure we can see clearly that the vertical profiles are contained in a cone defined by two straight lines. The vertical profiles for landing are similar. This behavior is mainly due to the different aircraft mass and performance and to the effect of the wind.

In a horizontal plan, we define a starting point A (x_A, y_A) and an ending point B (x_B, y_B). In a SID case, the starting point is at the runway threshold and the ending point is an exit point of a TMA. In a STAR case, the starting point is an entry point of a TMA, and the ending point is the Final Approach Fix (FAF). The horizontal route γ_H is a smooth mapping defined as:

$$\gamma_H : [0, 1] \to \mathbb{R}^2 \tag{1}$$

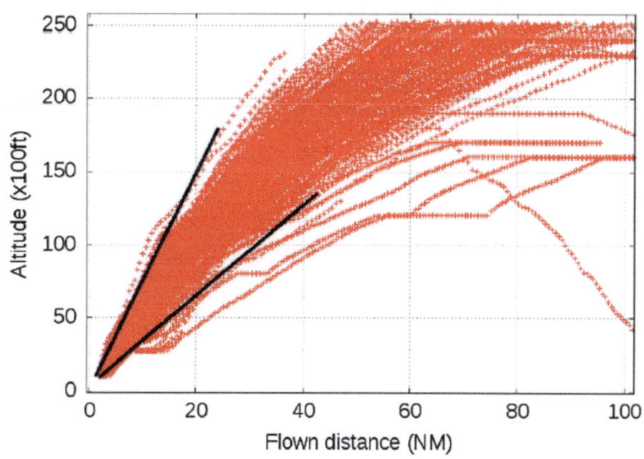

Fig. 1 Take-off profiles in CDG airport

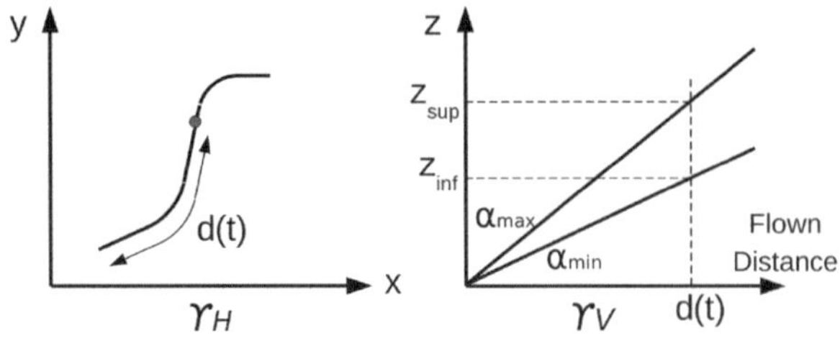

Fig. 2 An example of γ_H and γ_V

where $\gamma_H(0) = (x_A, y_A)$ and $\gamma_H(1) = (x_B, y_B)$. In a vertical plan, γ_V is defined as:

$$\gamma_V : \begin{array}{l} [0,1] \rightarrow I^{\mathbb{R}} \\ \quad t \quad \rightarrow [z_{inf}(d(t)), z_{sup}(d(t))] \end{array} \tag{2}$$

where $I^{\mathbb{R}}$ defines the set of intervals of $\mathbb{R}$, $d(t) = \int_0^t \|\gamma_H'(s)\|_2 \, \mathrm{d}s$ is the flown distance until t in horizontal plan, and $[z_{inf}(d), z_{sup}(d)]$ is the interval defined by the cross section of the cone at d. The vertical profile γ_V is associated with the horizontal curve γ_H through the flown distance d. Figure 2 illustrates an example of how γ_H is associated with γ_V in the case of a SID, where α_{min} and α_{max} are the minimum and maximum values of takeoff rate of aircraft on this route.

In the case of a SID, the starting point A is associated with an altitude z_A; the ending point B is associated with an altitude interval $[\underline{z_B}, \overline{z_B}]$ which indicates the altitude interval to exit TMA. Therefore, the boundary conditions are $\gamma_V(0) = [z_A, z_A]$ and $\gamma_V(1) \subset [\underline{z_B}, \overline{z_B}]$. Similarly, in the case of a STAR, the altitude interval of the starting point A is $[\underline{z_A}, \overline{z_A}]$ which indicates the interval to enter TMA; the altitude of the ending point B is z_B. The boundary conditions are $\gamma_V(0) = [\underline{z_A}, \overline{z_A}]$ and $\gamma_V(1) = [z_B, z_B]$. Note that, in order to guarantee that any SID reaches B within $[\underline{z_B}, \overline{z_B}]$, we set $\underline{z_B} = z_A + \alpha_{min} \cdot d(A, B)$, where α_{min} is the minimum taking-off slope and $d(A, B)$ is the Euclidean distance between the starting and ending points A and B. Similarly, to guarantee that any STAR reaches z_B at FAF, we set $\underline{z_A} = z_B + \alpha_{min} \cdot d(A, B)$ and $\overline{z_A} = z_B + \alpha_{max} \cdot d(A, B)$, where α_{min} (α_{max}) is the minimum (maximum) landing slope. A level flight can be imposed to ensure that $\overline{z_B}$ (respectively, z_B) is not exceeded in the case of a SID (respectively, STAR).

The obstacles, together with their protection areas, in number of $m \in N$, are modeled as cylinders in 3D, whose bases are parallel to the horizontal plan as presented in Fig. 3. Each cylinder $\Omega_i, i = 1, \cdots, m$ is defined by $(C_i(x_i, y_i), r_i, z_{i_{inf}}, z_{i_{sup}})$, where $C_i(x_i, y_i)$ and r_i are the center and the radius of the two bases, respectively; $z_{i_{inf}}$ and $z_{i_{sup}}$ are the altitude of the lower and upper bases. These obstacles are numbered in an increasing order of $length(A, Proj_{(AB)}C_i)$ where $Proj_{(AB)}C_i$ is the projection of C_i onto the line (AB). An illustration is presented in Fig. 4.

Let us define an *active* obstacle as an obstacle that is touched by a route and has to be avoided according to one of the following maneuvers: turn counterclockwise, turn clockwise, or impose a level flight. Each cylinder Ω_i is associated with two

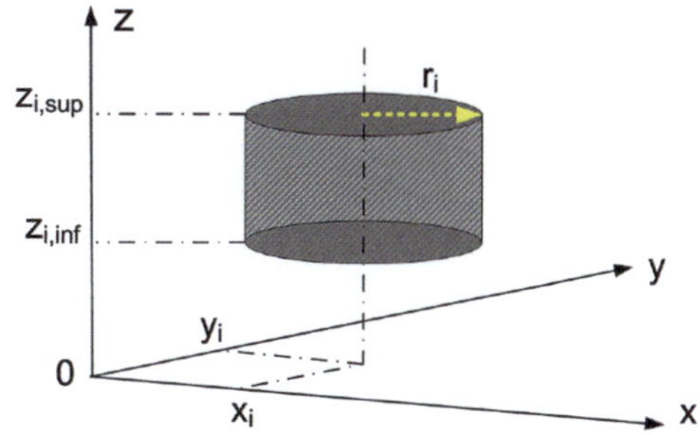

Fig. 3 Obstacle modelization

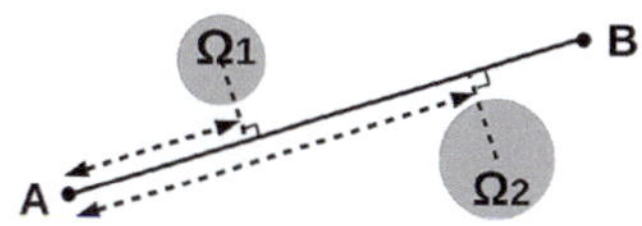

Fig. 4 Obstacles numbering

decision variables s_i and t_i: s_i defines whether Ω_i is active or not:

$$s_i = \begin{cases} 0, \text{ if } \Omega_i \text{ not active} \\ 1, \text{ if } \Omega_i \text{ active} \end{cases} \tag{3}$$

while t_i defines the ways an active obstacle Ω_i is avoided:

$$t_i = \begin{cases} 0, \text{ if turn counter-clockwise} \\ 1, \text{ if turn clockwise} \\ 2, \text{ if impose level flight} \end{cases} \tag{4}$$

Once the values of decision variables are chosen, the horizontal route is computed by connecting tangentially the successive active ($s_i = 1$) obstacles with $t_i = 0$ or $t_i = 1$ in the increasing order of their numbering. Active obstacles with $t_i = 0$ are bypassed counterclockwise and those with $t_i = 1$ are bypassed clockwise. This horizontal route is hence built piecewise: it is composed by $(\sum_{i=1}^{m} s_i - \sum_{i=1}^{m} \max(t_i - 1, 0) + 1)$ straight line segments and $(\sum_{i=1}^{m} s_i - \sum_{i=1}^{m} \max(t_i - 1, 0))$ arcs of circles. Routes are therefore constrained to lie on the border of obstacles. In order to ensure flyable routes, the radius of the arcs of circles is imposed to be at least equal to 3 Nm (FAA Orders 8260.54A and 8260.58). Note that these arcs can be followed using a type of performance-based navigation (PBN) named required navigation performance (RNP). Then, a vertical profile is associated with the horizontal route, taking into account α_{min}, α_{max} and imposing a level flight below (respectively, above) the active obstacle Ω_i in a SID (respectively, STAR) case when $t_i = 2$. If some active obstacles with $t_i = 2$ is not intersected by the cone associated with the horizontal route, then the route is unfeasible regarding our definition of "active obstacle." Note that the way of building a horizontal route simplifies the computation, but it does not necessarily lead to the shortest horizontal route between A and B.

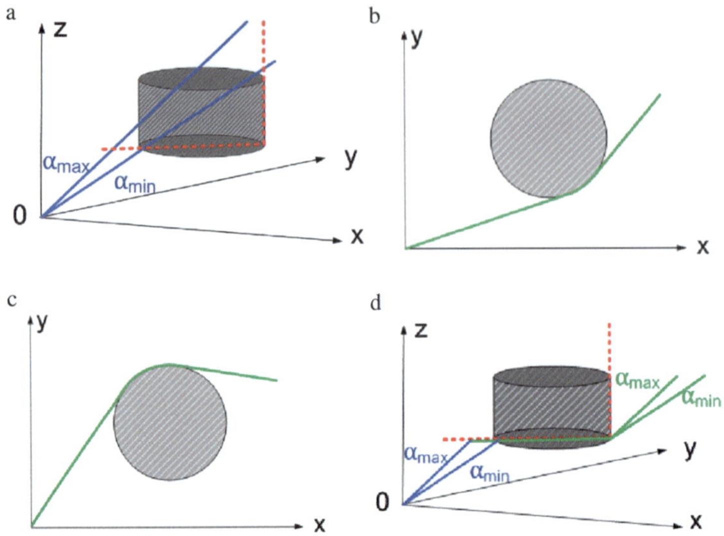

Fig. 5 The routes associated to different values of the decision variables. (**a**) $s_i = 0$, 3D view. (**b**) $(s_i, t_i) = (1, 0)$, 2D view. (**c**) $(s_i, t_i) = (1, 1)$, 2D view. (**d**) $(s_i, t_i) = (1, 2)$, 3D view

An illustration of different values of the decision variables for an example of a SID in a TMA with one obstacle is presented in Fig. 5. In Fig. 5a, the obstacle is not active, so $s_i = 0$. The horizontal route is a straight line segment connecting A and B. It is associated with a cone in the vertical plan. This route in the considered example is not a feasible one, because it intersects the obstacle. Then when the obstacle is active ($s_i = 1$), three possibilities are considered to avoid it (Fig. 5b–d): turn counterclockwise, turn clockwise, and impose a level flight under the obstacle at altitude $z_{i_{inf}}$, corresponding to $t_i = 0, 1, 2$, respectively.

Two examples of SIDs are presented in Fig. 6 showing how a route is computed, given the values of the decision variables, in the case with more than one obstacle. In example 1 (Fig. 6a, b), given $(s_1, t_1) = (1, 0), (s_2, t_2) = (1, 1)$, the horizontal route is composed by five parts: three segments and two arcs of circles. The three segments are used to connect tangentially the starting point to Ω_1, Ω_1 to Ω_2, and Ω_2 to the ending point, respectively. The two arcs are used to bypass Ω_1 counterclockwise and Ω_2 clockwise, respectively. In example 2 (Fig. 6c, d), given $(s_1, t_1) = (1, 2), (s_2, t_2) = (1, 1)$, the horizontal route is constructed by only bypassing Ω_2; thus, it is composed by two segments and one arc of circle. In vertical plan, when the route reaches the altitude of the lower basis of Ω_1, a level flight is imposed. The level flight ends at the flown distance where the horizontal route passes the boundary of Ω_1 and is no more intersected by Ω_1.

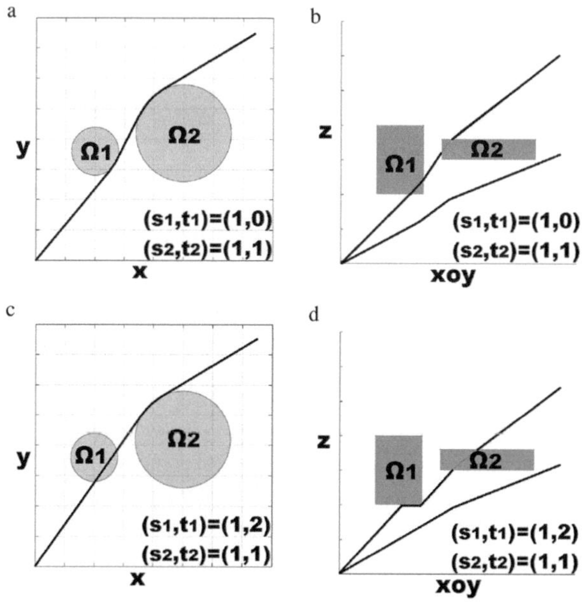

Fig. 6 Routes construction. (**a**) Example 1: Horizontal plan. (**b**) Example 1: Vertical plan. (**c**) Example 2: Horizontal plan. (**d**) Example 2: Vertical plan. Note that in (**b**) and (**d**), slopes appear discontinuous as an effect of a projection of a 3D image on a plan

Besides obstacle avoidance, further constraints are related to level flights. First, the number of level flights is bounded by a maximum number N_{max}, usually fixed to 2, for each route:

$$\sum_{i=1}^{m} \max(t_i - 1, 0) \leq N_{max} \tag{5}$$

Second, as the altitudes of imposed level flights have a direct impact on the noise pollution, a minimum altitude H_{min} for each level flight is defined. In practice, we impose the following constraints: in a SID case (respectively, a STAR case), for an obstacle Ω_i, if $z_{i_{inf}} < H_{min}$ (respectively, $z_{i_{sup}} < H_{min}$), then no level flight is imposed below (respectively, above) it; therefore, $t_i \in \{0, 1\}$.

Third, as to take into account the passengers' comfort, the length of each level flight should not be too short; a minimum length L_{min} for each level flight is imposed.

We minimize a weighted sum L_γ of the length of the route γ in the horizontal plan and the length related to the level flights. More precisely,

$$L_\gamma = c_1 \times \left(\int_0^1 \|\gamma_H'(t)\|_2 \, \mathrm{d}t \right) + c_2 \times L_{min} \times \sum_{i=1}^{m} \max(t_i - 1, 0) \tag{6}$$

The coefficients c_1 and c_2 are two penalty parameters; their values depend on the importance of the corresponding term. The obtained problem (denoted by $(\mathcal{P})$) is a combinatorial optimization problem. In the next section we explain the proposed solution approach for this problem.

3 Solution Approach: Branch and Bound

To solve the problem, we apply a branch and bound method. In [2], a path planning problem avoiding circular obstacles is studied in 2D. A branching strategy is proposed, where, for each obstacle, two branches are created depending on the clockwise or counterclockwise obstacle bypassing. We extend this branching strategy to take into account the specificity of our problem, where obstacles can be avoided also by imposing a level flight below (SID case) or above (STAR case) the obstacle. Our branching strategy is illustrated in Fig. 7. We start by setting Ω_i as active ($s_i = 1$) or not ($s_i = 0$); when it is active, we develop three branches in order to account for the three possibilities of avoiding it: counterclockwise ($t_i = 0$), clockwise ($t_i = 1$), or imposing a level flight ($t_i = 2$).

The lower bound for each subproblem is computed by generating the route which bypasses only the active obstacles and by calculating its length according to the objective function (6). The obtained lower bound is then used to identify whether a branch requires further subdivisions.

We present a step-by-step illustration (Fig. 8) to show how the branch and bound method works. The starting and ending points as well as two obstacles Ω_1, Ω_2 are presented in Fig. 8a. We take $c_1 = 1, c_2 = 1$, that is to say we penalize the length of level flights in the objective function.

Step 1: We develop four branches on Ω_1. We start by deviating the route counterclockwise, and we obtain a route that does not intersect Ω_2. The lower bound in this case is 100,320 m. Besides, the value of the objective function associated with solution $(s_1, t_1) = (1, 0)$, $s_2 = 0$ is equal to the lower bound. Therefore, no further exploration is needed.

Step 2: Another branch on Ω_1 is developed by deviating the direct route clockwise around Ω_1. The length of this horizontal route is the lower bound of this subproblem; the value is greater than the current best value. There is no

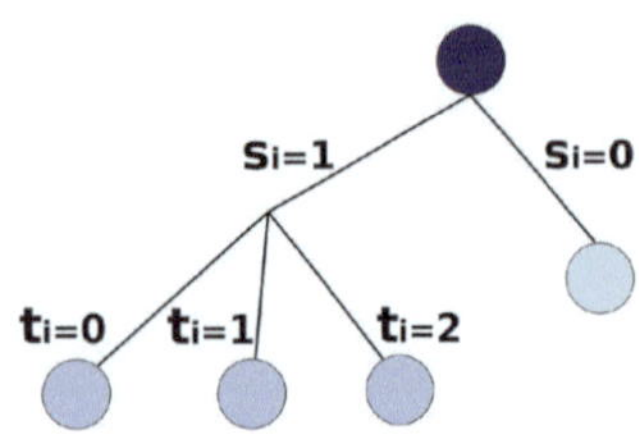

Fig. 7 Branch and bound branching strategy

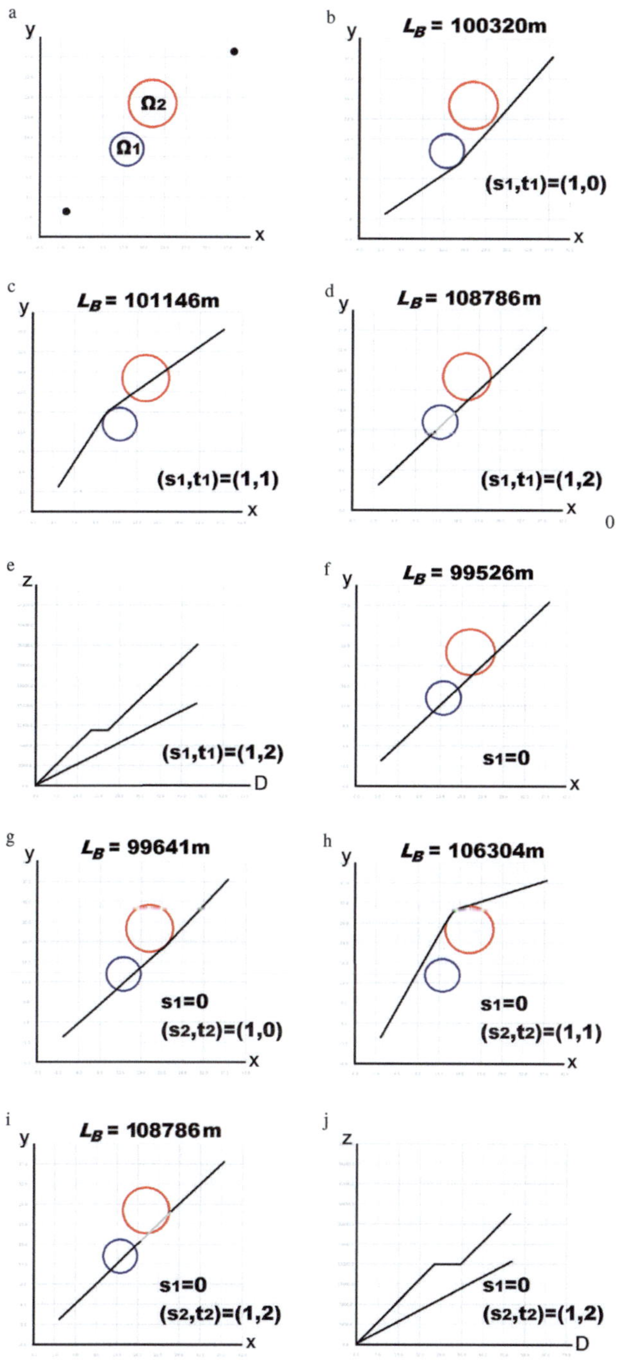

Fig. 8 Branch and bound illustration. (**a**) Step 0. (**b**) Step 1. (**c**) Step 2. (**d**) Step 3, horizontal plan. (**e**) Step 3, vertical plan. (**f**) Step 4. (**g**) Step 5. (**h**) Step 6. (**i**) Step 7, horizontal plan. (**j**) Step 7, vertical plan

possibility to get a better solution by further branching on Ω_2; therefore, we cut this branch.

Step 3: The third branch on Ω_1 is obtained by imposing a level flight, as shown in Fig. 8d, e. The lower bound is greater than the current best value, so the branch is cut.

Step 4: The last branch on Ω_1 is with $s_1 = 0$. The route intersects Ω_2, and its lower bound, which corresponds to the length of the direct route, is less than the current best length; thus, four branches on Ω_2 are developed. In the case $s_2 = 0$, the route is not feasible and thus not accepted.

Step 5: By branching counterclockwise around Ω_2, the obtained route is still intersecting Ω_1; thus, it is not accepted.

Step 6: By branching clockwise around Ω_2, we obtain a feasible route with length greater than the current best value, so it is not accepted.

Step 7: The last branch is obtained by imposing a level flight under Ω_2 as shown in Fig. 8i, j. This route is still encountered by Ω_1, so it is not accepted.

All the possible branches are considered. The best distance is 100,320 m by taking $(s_1, t_1) = (1, 0), s_2 = 0$.

The maximum size of the search space of our problem is 4^m, where m is the number of obstacles. Indeed we have four possibilities to deal with each obstacle. The size of the search space can be reduced significantly by applying the preprocessing techniques developed in [4]. In fact, the authors prove firstly that the shortest path connecting two points and avoiding circular obstacles must lie into an ellipse containing these two points. Moreover, they prove that the shortest path lies in a convex hull of a few circular obstacles around the line segment connecting the starting and ending points. We apply these preprocessing techniques to reduce the number of considered obstacles before applying our approach. These filters still hold when considering cylindrical obstacles. A 3D feasible route can be indeed built based on the 2D shortest path avoiding the obstacle projections on the plan. Some simulation results with and without the preprocessing techniques are presented in Sect. 4.

4 Simulation Results

In this section we present two simulation examples. The first one is the case of an SID, and the second one is the case of a STAR. Tests were run on a Linux platform with a 2.4 GHz processor and 8 GB RAM.

Different strategies are applied to choose the next subproblem to solve and the next obstacle to branch on. The simulation results presented in this section use the best lower bound (BLB) for selecting the next subproblem, and use the first intersected obstacle (FIO) for selecting the next obstacle to branch on. In fact, the combination of strategies "BLB+FIO" gives the minimum computing time together with the minimum number of iterations in most of the tests.

Table 1 Example 1: characteristics of nine obstacles

(x_i, y_i) (m)	r_i (m)	$(z_{i_{inf}}, z_{i_{sup}})$ (m)
(7408,14816)	3704	(701, 2377)
(9260,35100)	111,12	(884, 1707)
(9260,57412)	4630	(2286, 4389)
(29632,25928)	9260	(2682, 3109)
(33336,48152)	5556	(2835, 5425)
(42596,14816)	5556	(2286, 4389)
(48152,51856)	7408	(3536, 6828)
(61116,29632)	9260	(3048, 6706)
(70376,51826)	9260	(4154, 8022)

Table 2 Example 1: simulation results

	No pre-processing		With pre-processing		
(c_1, c_2)	Time (s)	Iterations	Time (s)	Iterations	L_γ (m)
$(1, 0)$	0.47	254	0.09	46	100,082
$(1, 1)$	0.57	307	0.20	115	102,478

In the first example, the input data are (unit in meter):

- Starting point A, $(x_A, y_A, z_A) = (0, 0, 0)$
- Ending point B, $(x_B, y_B, [\underline{z_B}, \overline{z_B}]) = (70{,}376, 70{,}376, [4777, 9754])$
- $\alpha_{min} = 4.8\%, \alpha_{max} = 9.2\%$
- $N_{max} = 2, L_{min} = 9260, H_{min} = 914$

There are nine obstacles. Table 1 gives the center (x_i, y_i), radius r_i, and altitudes of the two bases $(z_{i_{inf}}, z_{i_{sup}})$ of obstacle $\Omega_i, i = 1, \cdots, 9$. The unit is in meter. The simulation results are presented in Table 2 and Fig. 9. We carried out two tests with different values of c_1, c_2: the first one with $c_1 - 1, c_2 - 0$ and the second one with $c_1 = 1, c_2 = 1$. When $c_1 = 1, c_2 = 0$, the length of level flights is not penalized in the objective function; thus, Fig. 9a, b shows that the optimal route is obtained with a level flight to avoid the obstacle with the center coordinates (29,632, 25,928). However, when $c_1 = 1, c_2 = 1$, the length of level flight is penalized in the objective function, and as a result a counterclockwise turn is made to avoid the mentioned obstacle instead of using a level flight, as shown in Fig. 9c, d. Moreover, thanks to the ellipse and convex hull filters, which reduces the number of the potential obstacles from 9 to 5, the computation time and the number of iterations are reduced effectively.

In the second example, the input data are (unit in meter):

- Starting point A, $(x_A, y_A, [\underline{z_A}, \overline{z_A}]) = (0, 0, [4454, 7722])$
- Ending point B, $(x_B, y_B, z_B) = (96{,}304, 96{,}304, 914)$
- $\alpha_{min} = 2.6\%, \alpha_{max} = 5\%$
- $N_{max} = 2, L_{min} = 9260, H_{min} = 914$

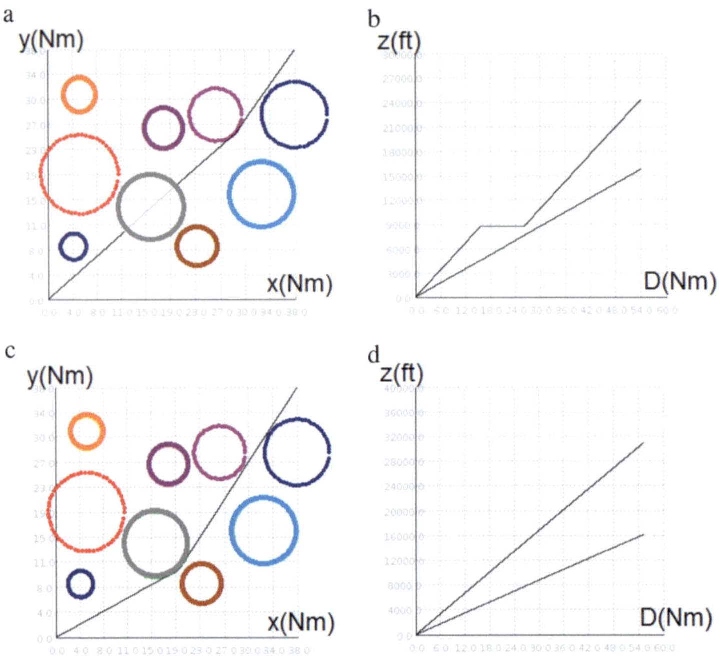

Fig. 9 Example 1: simulation results. (**a**) $\gamma_H, c_1 = 1, c_2 = 0$. (**b**) $\gamma_V, c_1 = 1, c_2 = 0$. (**c**) $\gamma_H, c_1 = 1, c_2 = 1$. (**d**) $\gamma_V, c_1 = 1, c_2 = 1$

Table 3 Example 2: characteristics of seven obstacles

(x_i, y_i) (m)	r_i (m)	$(z_{i_{inf}}, z_{i_{sup}})$ (m)
(22224,22224)	9260	(0,5486)
(27780,74080)	148,16	(0,3048)
(48152,96304)	7408	(0,4572)
(59264,51856)	129,64	(0,4572)
(72228,24076)	9260	(0,4572)
(77784,81488)	148,16	(0,3048)
(96304,48152)	129,64	(0,4572)

Table 4 Example 2: simulation results

(c_1, c_2)	No pre-processing		With pre-processing		L_γ (m)
	Time (s)	Iterations	Time (s)	Iterations	
(1, 0)	0.64	349	0.09	45	142,230
(1, 1)	0.36	188	0.09	45	142,230

There are seven obstacles defined by the parameters presented in Table 3. The simulation results are presented in Table 4 and Fig. 10. Note that the optimal route in the case with $c_1 = 1, c_2 = 0$ is the same as the one in the case with $c_1 = 1, c_2 = 1$. Even though by taking $c_2 = 0$, level flights are not penalized in the objective

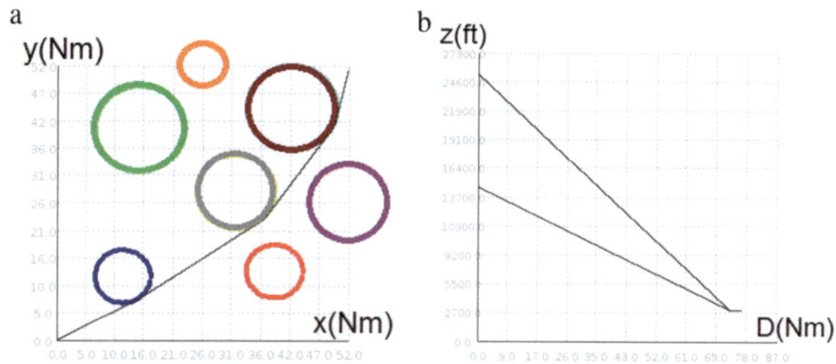

Fig. 10 Example 2: simulation results. (**a**) $\gamma_H, c_1 = 1, c_2 = 0$ or $c_1 = 1, c_2 = 1$. (**b**) $\gamma_V, c_1 = 1, c_2 = 0$ or $c_1 = 1, c_2 = 1$

function, no level flight is imposed in the optimal solution. The reason is that the routes with level flights do not offer a better value of the objective function or do not satisfy the constraints that mentioned before. By applying the preprocessing techniques, the number of the potential obstacles is reduced from 7 to 3; therefore, the computation time and the number of iterations are reduced significantly.

5 Conclusion and Perspectives

In this paper, we introduce a methodology for generating a 3D route in TMA at strategic level, performed by a deterministic global optimization approach. The route is represented by a horizontal curve associated with a cone in the vertical plan. We develop three ways to avoid an obstacle: turn clockwise or counterclockwise (2D) and level flight (3D) which correspond to branching strategies in a branch and bound tailored on the problem. By setting appropriately the penalization coefficients, it is possible to obtain continuous and smooth routes which are available for Continuous Climb Operations (CCO) and Continuous Descent Operations (CDO). To summarize, this approach can be regarded as a decision support tool for the designing of SID/STARs.

In future work, we will consider the problem of designing several routes taking into account the avoidance of obstacles and the separation between them. One possible approach is to decompose the problem in three parts: (a) generating each route separately by applying the branch and bound method presented in this paper, (b) detecting the conflicts between the generated routes, and (c) eliminating the conflicts by adding fictitious cylinder obstacles at the position of the conflicts.

Acknowledgements This work has been partially supported by Civil Aviation University of China, by National Natural Science Foundation of China (NNSFC) through grant NNSFC 61201085, and by French National Research Agency (ANR) through grant ANR 12-JS02-009-01

ATOMIC. The authors would like to thank Serge Roux from the French Civil Aviation University (ENAC), for the interesting discussions and useful comments.

References

1. Delahaye, D., Puechmorel, S., Tsiotras, P., Feron, E.: Mathematical models for aircraft trajectory design: A survey. Air Traffic Management and Systems Lectures Notes in Electrical Engineering. **290**, 205–247 (2014)
2. Eele, A., Richards, A.: Path planning with avoidance using nonlinear branch and bound optimization. J. Guid. Control. Dynam. **32**, 384–394 (2009)
3. Gallina, P., Gasparetto, A.: A technique to analytically formulate and to solve the 2-dimensional constrained trajectory planning problem for a mobile robot. J. Intell. Robotics Syst. **27**, 237–262 (2000)
4. Kim, D.S., Yu, K., Cho, Y., Kim, D., Yap, C.: Shortest paths for disc obstacles. Lect. Notes Comput. Sci. **3045**, 62–70 (2004)
5. Lozano-Pérez, T., Wesley, M.A.: An algorithm for planning collison-free paths among polyhedral obstacles. Commun. ACM. **22**, 560–570 (1979)
6. Pfeil, D.M.: Optimization of airport terminal-area air traffic operations under uncertain weather conditions. PhD thesis, Massachusetts Institute of Technology, 2011
7. Pocchiola, M., Vegter, G.: Minimal tangent visibility graphs. Comput. Geom. Theory Appl. **6**, 303–314 (1996)
8. Rohnert, H.: Shortest paths in the plane with convex polygonal obstacles. Inf. Process. Lett. **23**, 71–76 (1986)
9. Souissi, O. et al.: Path planning: A 2013 survey. In: Proceedings of the 2013 International Conference on Industrial Engineering and Systems Management, pp. 1–8. Rabat (2013)
10. Storer, J.A., Reif, J.H.: Shortest paths in the plane with polygonal obstacles. J. ACM. **41**, 982–1012 (1994)

Arrival Time Assignment by Dynamic Programming Optimization

Haruki Matsuda, Akinori Harada, Tomoyuki Kozuka, Yoshikazu Miyazawa, and Navinda Kithmal Wickramasinghe

Abstract Japanese airspace capacity must expand in order to accommodate the increased air traffic expected in the near future. Efficient air congestion management is a promising approach for achieving this goal. Arrival management for inbound flights to Tokyo International Airport, the busiest airport in Japan, is considered to be the most demanding challenge for efficient air congestion management. In this paper, a concept of arrival management based on multiple aircraft trajectory optimization is proposed and examined using the actual flight track data. At first, free-flight trajectory optimization is applied to the inbound flights landing on one of the two runways to simulate the most efficient ideal flights. Next, time separation constraints at a merging point on the boundary of the terminal area are imposed in order to avoid conflicts among the aircraft in the terminal area. As a result of the optimization, optimal sequencing and flight time adjustments are generated. Benefits of the proposed concept are evaluated by comparing the optimal trajectories with the corresponding actual flight trajectories. Dynamic programming is used for the optimization of each flight trajectory and scheduling of the arrival times. The obtained results reveal that in addition to safe arrival time separation, trajectory optimization with arrival time assignment produces substantial benefits in terms of fuel consumption and flight time.

Keywords Arrival management • Arrival time assignment • Dynamic programming • Trajectory optimization

---

H. Matsuda (✉)
Department of Aeronautics and Astronautics, Graduate School of Engineering, Kyushu University, Fukuoka, Japan

Kawasaki Heavy Industries, Ltd., Tokyo, Japan
e-mail: harukimatsuda1108@gmail.com

A. Harada • T. Kozuka • Y. Miyazawa • N.K. Wickramasinghe
Department of Aeronautics and Astronautics, Graduate School of Engineering, Kyushu University, Fukuoka, Japan

Electronic Navigation Research Institute, *Air Traffic Management and Systems II*, Lecture Notes in Electrical Engineering 420, DOI 10.1007/978-4-431-56423-2_10

1 Introduction

Japanese airspace capacity must increase to handle the continuously growing air traffic. Management of a more crowded airspace is difficult under the current system owing to the increased economic costs caused by the required detours and a higher danger of collision. Additionally, fuel prices are considerably high, and there is a rising demand for the reduction of environmental problems in the airline industry. Consequently, the airlines are making unprecedented efforts to reduce flight costs [1, 2]. In response to these demands, the Collaborative Actions for Renovation of Air Traffic System (CARATS) was formulated in Japan; this is the counterpart of the program for the Next Generation Air Transportation System (NextGen) in the USA and the Single European Sky ATM Research Program (SESAR) in Europe [3–5].

Currently, the air traffic management (ATM) system uses predetermined routes and divides the airspace into sectors that are managed by controllers in charge. However, the application of the ATM system in Japan results in frequent detours with delays in particular airspaces or routes especially at major airports. Using trajectory-based operation (TBO) approach mentioned in CARATS instead of the current ATM system is a potential solution for the poor ability of the ATM system to cope with increasing air traffic flow. According to this operational concept, all flights from departure to arrival are integrally managed as trajectories including time. Therefore, the TBO is considered to be an improvement on the conventional operations and can be applied to optimize the entire Japanese Flight Information Region (FIR) through the ATM system. Furthermore, the TBO allows preferred route selection based on each aircraft's operational performance (commonly referred to as the free-flight concept).

The potential benefits of realizing such a system were evaluated in previous studies [6–9]. Although this analysis did not consider realistic restraints such as the actual flight rules, it revealed that trajectory optimization offers significant benefits particularly for the current airliner flights descent phases.

The entire airspace capacity is constrained by the throughput of the most congested airspace. The trajectory-based approach enables effective management of flights arriving at a congested airport and leads to an increasing throughput at the congested airspace. This eliminates the bottleneck for the expansion of airspace capacity and allows a significant improvement of flight's arriving efficiency because most arriving flights are in their descending phases.

Furthermore, arrival management is important for realizing the free-flight concept. If several aircraft are permitted to freely arrive at the same runway, there is a greater possibility of conflict. In this study, we attempt to assign arrival times to each aircraft for scheduling their arrivals with appropriate time separations.

The potential benefits of arrival time assignment for Tokyo International Airport, the most congested airport in Japan, were evaluated in a previous study [10]. However, this analysis requires the start times of all flights. There are considerably large differences in the start time of the flights having different flight ranges even if they have the same arrival time. Thus, it is not realistic to assign arrival times immediately after estimating them for flights with different start times. Therefore, the method used in the previous study [10] produces unrealistic results in terms of actual operation. In the present study, the arrival time assignment is performed dynamically. Furthermore, we show that both ensuring safe arrival time separations and the realization of the benefits of this approach can be simultaneously achieved by trajectory optimization with arrival time assignment for the aircraft arriving at Tokyo International Airport.

2 Operational Efficiency Evaluation

Flight state estimation and trajectory optimization are performed to evaluate the efficiency of actual airliner operation. The actual airliner flight states are estimated using the flight track data. Thus, fuel consumption and flight time required for the actual flights can be obtained and used as evaluation criteria. On the other hand, flight trajectory optimization is used to derive the flight states for the flights having superior performance efficiency. As a result, we can evaluate the operational efficiency of the actual flights by comparing the optimal flight states to the estimated flight states.

Figure 1 illustrates the flow chart of the operational efficiency evaluation procedure. First, the position change rate of the aircraft (the ground speed) is derived from the flight track data. Air data including airspeed and Mach number are also estimated using the meteorological information, such as temperature, barometric pressure, wind direction, and wind speed. Moreover, an aircraft performance model (APM) is used for estimating the fuel flow, which is then integrated over time to obtain the total fuel consumption.

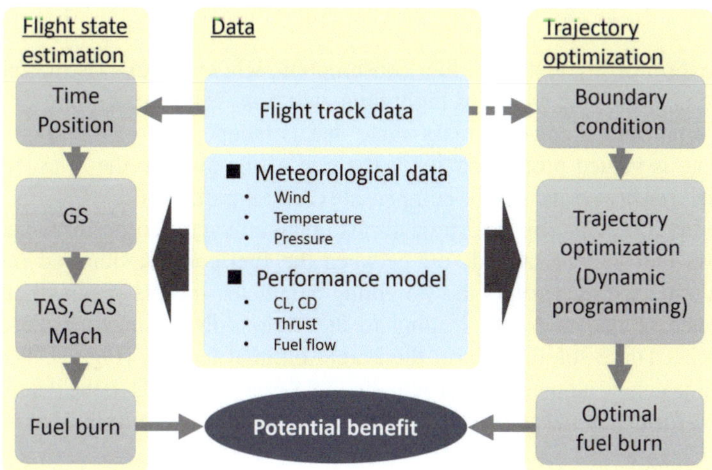

Fig. 1 Analysis of operational efficiency

Flight trajectory optimization is performed using the initial and final positions and estimated airspeed as the boundary conditions. The meteorological information and the APM used here are the same as those used in the flight state estimation in order to reduce potential errors inherent in the meteorological data and the APM. Consequently, the obtained fuel consumption and flight time are optimal with respect to the evaluation criteria, and the operational efficiency evaluation of the actual flights is performed assuming that these trajectories obtained by the optimization are the optimal flight states in terms of operational efficiency. In this study, we use the meteorological grid point value (GPV) data released by the Japan Meteorological Agency. The Base of Aircraft Data (BADA) Revision 3.11 [11] developed and maintained by EUROCONTROL is used as the APM. It was revealed that both the data and model exhibit sufficient accuracy for their application to the fuel consumption estimation or operational efficiency evaluation by comparison to the quick access recorder data stored by the airlines [12, 13].

3 Trajectory Optimization and Arrival Time Assignment

If multiple aircraft freely landing at the same runway choose optimal paths for which the time separations are not explicitly constrained, conflicts can occur at the terminal merging point. This is a major issue toward the realization of an air traffic system based on the free-flight concept. Here, we assign appropriate arrival times in order to resolve this issue.

The trajectory optimization of the kth aircraft is performed using the performance index, J_k, which is defined as follows:

$$J_k = \int_{t_{0k}}^{t_\mathrm{f}} \mu_k(t)\mathrm{d}t + \frac{m_k}{m_0} a_k \left(t_\mathrm{f} - t_{0k}\right) \tag{1}$$

The first term of Eq. (1) is the fuel consumption, where μ_k(kg/s) is the fuel flow. The second term is the weighted flight time, where m_k is the mass of the kth aircraft and m_0 represents a constant mass value that is independent of the aircraft type. Since time is valued proportionally to the size of the aircraft, the time-weighting parameter, $(m_k/m_0)a_k$, is used to compensate each aircraft's mass.

Examination of the optimal trajectories obtained using the performance index (1) shows that there is a trade-off between the fuel consumption and the flight time. In other words, varying the weighting parameter, a corresponds to changing the balance of this trade-off, leading to the various flight states of the optimal trajectories. These solutions form the Pareto optimal frontier (Fig. 2). While the arrival time can be adjusted by varying the flight time, this leads to the deterioration in the performance index values. The total penalty owing to time assignment is

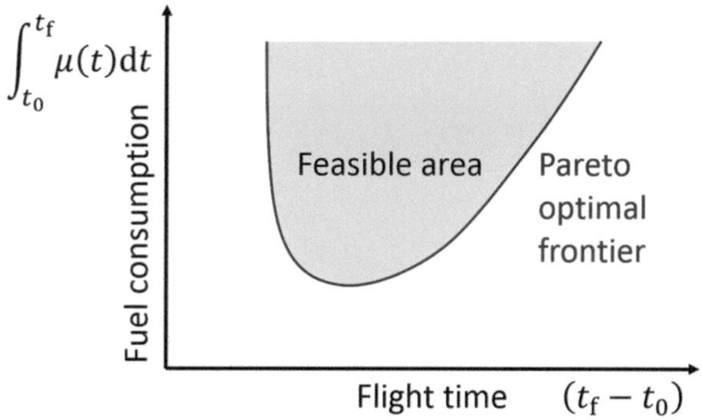

Fig. 2 Optimal fuel consumption and flight time

expected to be minimized, as suggested by the total performance indices given by Eqs. (2) and (3).

$$J_1^* = \sum_k J_k = \sum_k \left\{ \int_{t_{0k}}^{t_{fk}} \mu_k(t)\mathrm{d}t + \frac{m_k}{m_0} a_k \left(t_{fk} - t_{0k}\right) \right\} \tag{2}$$

$$J_2^* = \sum_k \frac{m_0}{m_k} J_k = \sum_k \left\{ \frac{m_0}{m_k} \int_{t_{0k}}^{t_{fk}} \mu_k(t)\mathrm{d}t + a_k \left(t_{fk} - t_{0k}\right) \right\} \tag{3}$$

Because various types of aircraft are assessed as arrival flights in the same framework, the arrival time adjustment tends to become smaller for larger aircraft. In this case, Eq. (2) provides the performance index. The performance index, J_2^*, in Eq. (3) is compensated for the aircraft's mass to prevent the underestimation of the arrival time adjustment. In Eq. (3), J_2^* is adopted as the performance index for arrival time optimization. The inequality constraint in Eq. (4) is considered to ensure the appropriate arrival time separations.

$$\left|t_{fk} - t_{fl}\right| > t_{\mathrm{min,separation}} \tag{4}$$

for any k and l, $k \neq l$.

The problem discussed in this study is called a bi-level programming problem, where the constraint condition of the arrival time optimization problem that is an upper problem includes other optimization problem for flight trajectory that is a lower problem. In this study, a set of solutions satisfying the constraint condition, i.e., optimal solutions for the flight trajectory optimization problem, is generated in advance, and then, solutions for the arrival time optimization problem are obtained as the combination of optimal flight trajectories. Dynamic programming is used for both optimization problems in this study. Figure 3 illustrates the conceptual scheme of the grid used for the arrival time optimization calculations. When the arrival sequence is defined as an independent variable, the aircraft ID and arrival time are optimized.

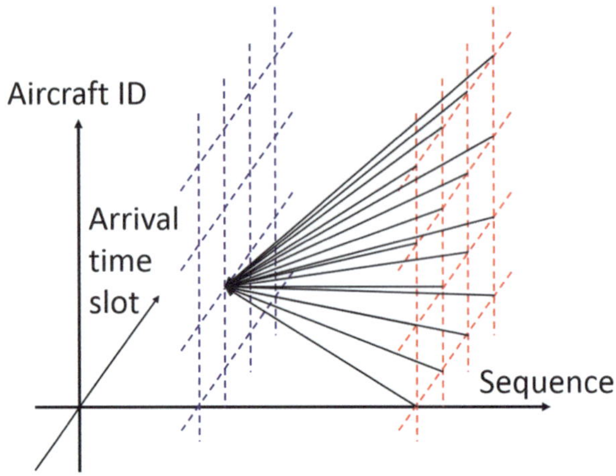

Fig. 3 Arrival time optimization by dynamic programming

4 Application to Tokyo International Airport

Tokyo International Airport (ICAO airport code: RJTT) manages the highest number of passengers in Japan and handles air traffic control operations with four runways. The RWY34L is used exclusively for landings during north wind operations, and it is not required to take conflicts with departing aircraft into consideration. This study focuses on the aircraft arriving at the RWY34L.

The track data used in this study are taken from the CARATS Open Data, which are released by the Japan Civil Aviation Bureau for promoting research and development in the field of air traffic management. These data are obtained from the radar data recorded by the surveillance radar for en route. The data comprises the virtual flight number, time, latitude, longitude, and barometric pressure related to the respective aircraft.

The flights toward Tokyo International Airport for 16 h from 7 a.m. to 11 p.m. on May 9, 2012, were analyzed, while the phases of takeoff and landing are eliminated. The initial point recorded at the Fukuoka FIR for each international flight is considered as the initial data point, and for domestic flights, the data point at approximately 10,000 ft in the climb phase is taken as the initial data point for the analysis. Additionally, the waypoint ADDUM, the initial fix on the standard terminal arrival route (STAR), is assumed to be the merging point in the terminal, which is the final point in the analysis. Figure 4 illustrates the characteristics of the airspace that is being modeled. However, the arrival time separations of actual flights are evaluated at the fix called ARLON that is located closer to the runway than ADDUM because some arriving flights are actually merged at this point by the current air traffic control system. The actual track chart and

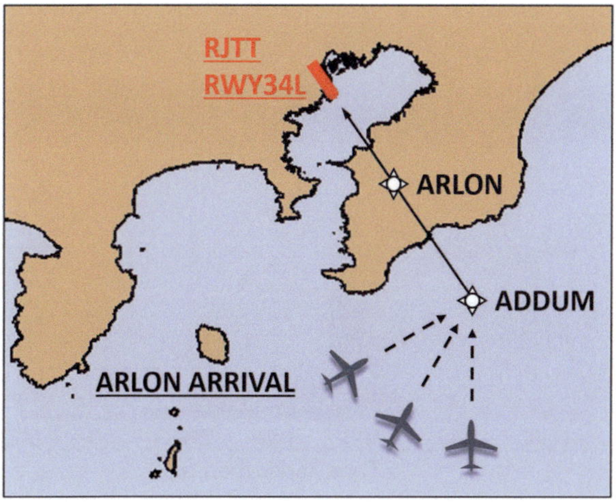

Fig. 4 Characteristics of the airspace

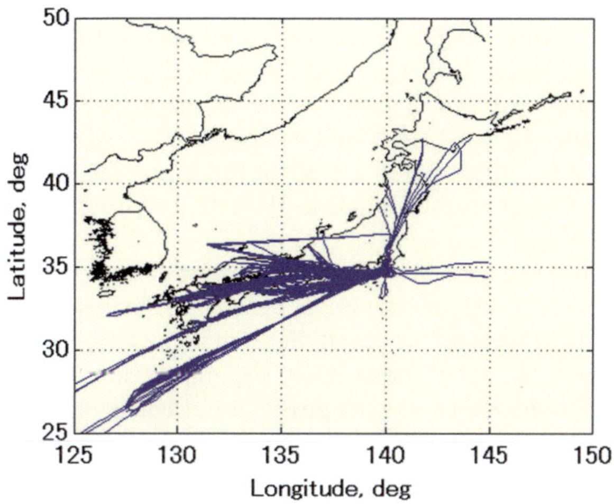

Fig. 5 Flight track (actual)

distribution of arrival time separations of the target flights are shown in Figs. 5 and 6, respectively.

We assume that the target flights are freely optimized without arrival time assignment using the performance index shown in Eq. (1). The parameters used for the optimization are given as follows.

$$a_k = 0.5\ \text{kg/s},\ \ m_0 = 208700\ \text{kg}$$

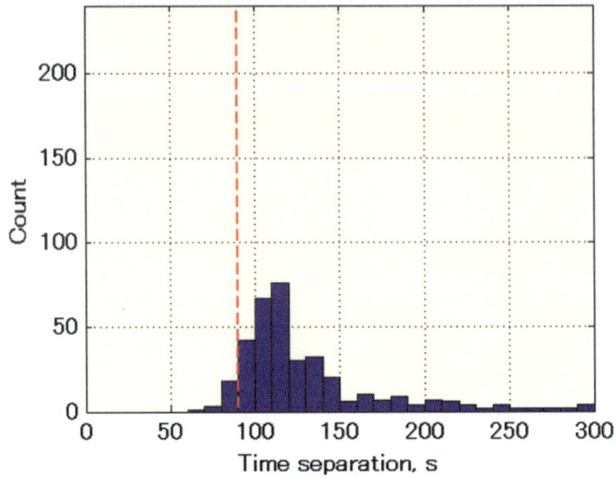

Fig. 6 Arrival time separations (actual)

This setting of the time-weighting parameter considers fuel consumption to be a relatively more important efficiency criterion than the flight time. Additionally, each aircraft mass is given as a 10 % deduction of the reference mass mentioned in BADA in order to consider Japanese short-haul domestic flights. Although the boundary conditions are set to correspond to the actual flights, for the terminal conditions, we generally set the barometric altitude to 10,000 ft and the calibrated airspeed to 230 kt at ADDUM.

In this study, two cases are presented for evaluating the potential benefits of flight trajectory optimization and arrival time assignment. Although optimal flights are compared with the corresponding actual flights in both cases, there is a significant difference between the cases. In case (a), the trajectories are freely optimized without any constraint so that there is possibility of conflicts at the final point in this case. To resolve this problem, in case (b), arrival time adjustment is performed with safe arrival time separation given as the follows.

$$t_{\text{min,separation}} = 90 \ \ s$$

The arrival time assignment is performed using the following procedure:

(a) The aircraft within the 60-min range, i.e., the range corresponding to 60-min flight to the merging point at terminal, are monitored. The arrival times are assigned to prevent any conflicts between the arrival times of the monitored aircraft.
(b) The arrival times of the aircraft within the 15-min range, corresponding to the 15-min flight to the merging point, are fixed. The arrival times of the aircraft beyond the 15-min range are adjusted to prevent any conflicts with the aircraft for which the arrival times have already been fixed.

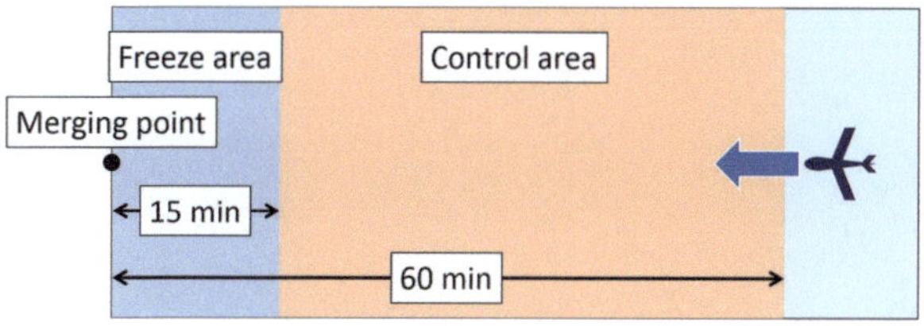

Fig. 7 Procedure for arrival time assignment

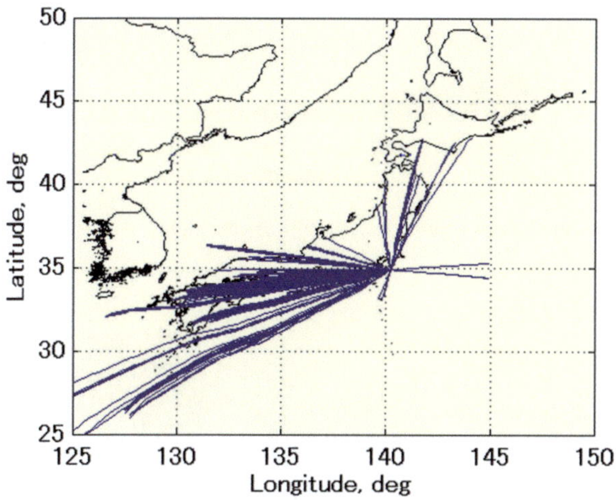

Fig. 8 Flight track (optimal, without time constraint)

Figure 7 schematically illustrates this procedure. The time step of the calculation is set to 5 min, and the arrival time assignment is performed every 5 min.

5 Results

The tracking chart and distribution of arrival time separations obtained by the free optimization in the case (a) are shown in Figs. 8 and 9, respectively. Moreover, Fig. 10 illustrates the differences between the optimal trajectories and the actual flights with respect to fuel consumption and flight time that are the criteria for the evaluation of the potential benefits. Furthermore, the averages of differences and their percent values are shown in Table 1. The flight tracks tend to choose more straight paths than those of the actual flights, and we can see that approximately a 10 % reduction in fuel consumption is achieved. However, safe arrival time separations between the arriving flights are not achieved.

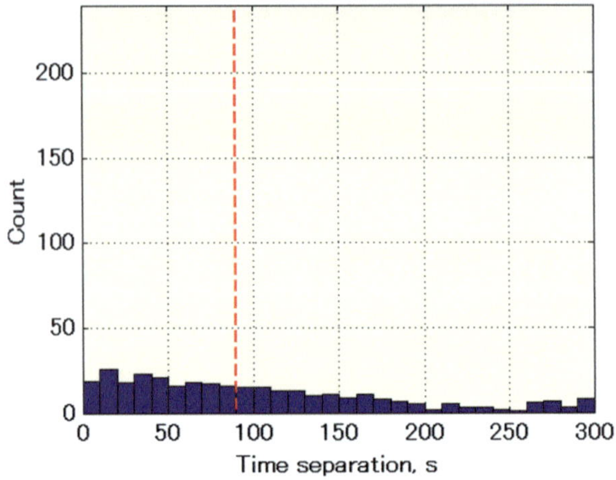

Fig. 9 Arrival time separations (optimal, without time constraint)

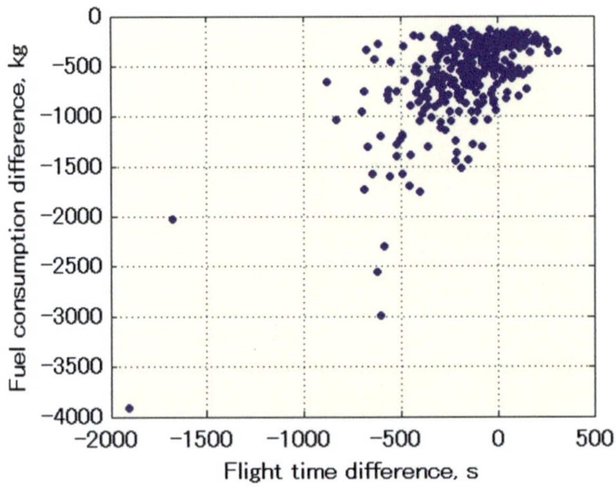

Fig. 10 Potential benefits by trajectory optimization (without time constraint)

Table 1 Average of potential benefits (without time constraint)

	Difference	Percentage
Fuel consumption	−559 kg	−13.24 %
Flight time	−149 s	−4.00 %

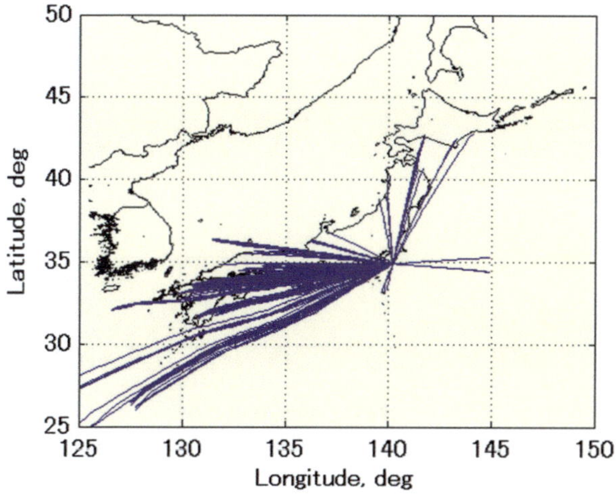

Fig. 11 Flight track (optimal, with time constraint)

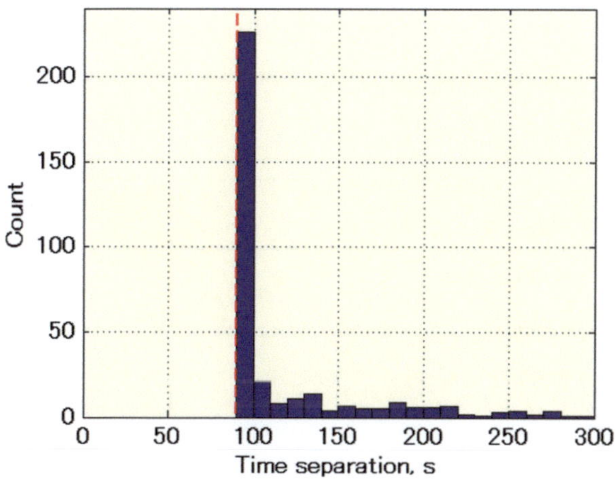

Fig. 12 Arrival time separations (optimal, with time constraint)

Figures 11 and 12 show the track chart and the distribution, respectively, of arrival time separations obtained for case (b). The arrival time separations are also plotted versus time in Fig. 13. The horizontal red lines in this figure indicate the boundary of safe time separation. Figure 13 shows that safe values of arrival time separations are preserved with almost direct routes. Additionally, the potential benefits obtained in this case are shown in Fig. 14 together with the potential benefits obtained by freely optimizing the flight trajectory without arrival time assignment.

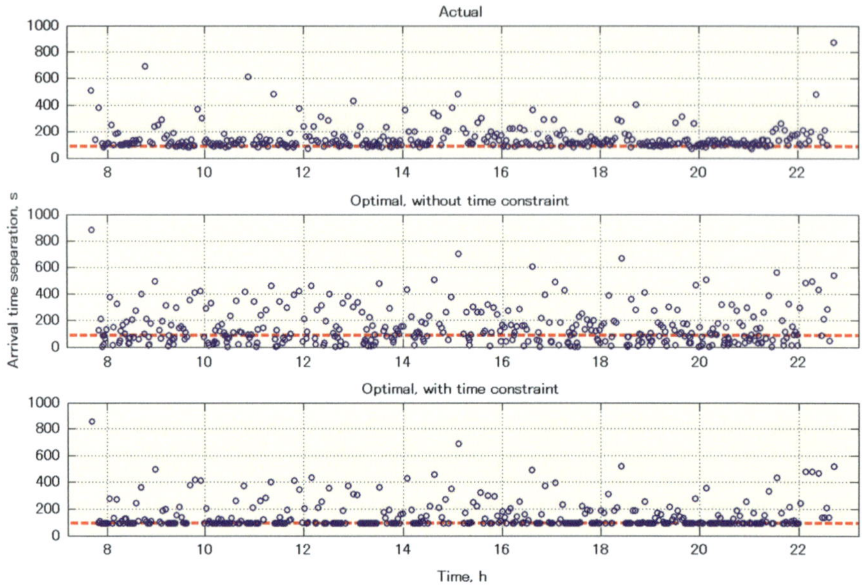

Fig. 13 Time history of arrival time separations

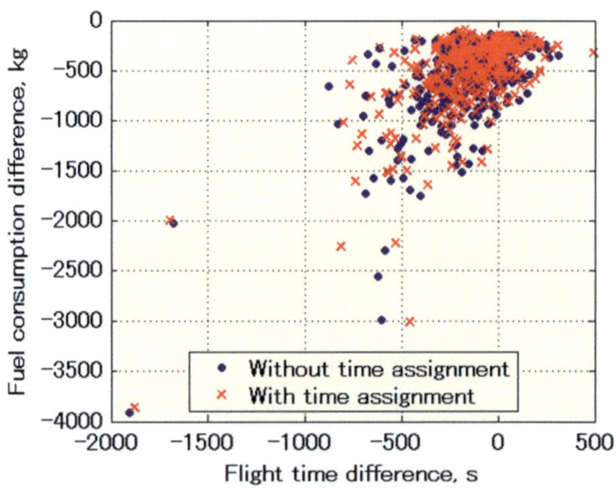

Fig. 14 Potential benefits by trajectory optimization

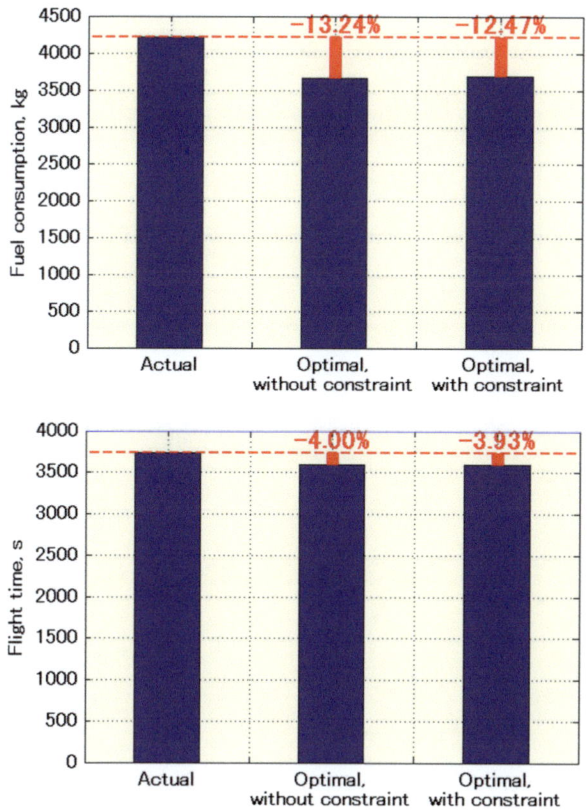

Fig. 15 Average of fuel consumption and flight time

Table 2 Average of fuel consumption and flight time for 375 flights

	Actual flight	Optimal trajectories (difference, ratio relative to actual flight)				
		(a) Without time constraint		(b) With time constraint		(b)–(a)
Fuel burn (kg)	4217	3658	(−559, −13.24 %)	3691	(−526, −12.47 %)	+33
Flight time (s)	3741	3592	(−149, −4.00 %)	3594	(−147, −3.93 %)	+2

Further, Fig. 15 illustrates the average of the difference between the actual flights and the optimal flights in terms of fuel consumption and flight time, and Table 2 shows the numeric values. There is no significant trend in the potential benefit data; irrespective of whether the arrival times are assigned, high benefits can potentially be obtained in both cases.

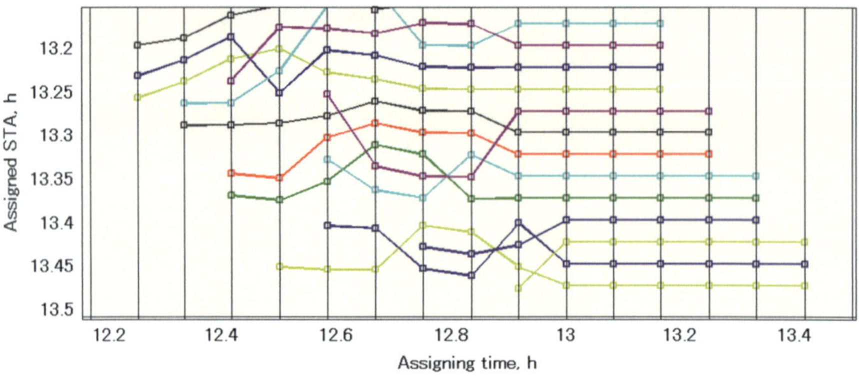

Fig. 16 Time history of arrival time assignment for 1 h (STA, scheduled time of arrival)

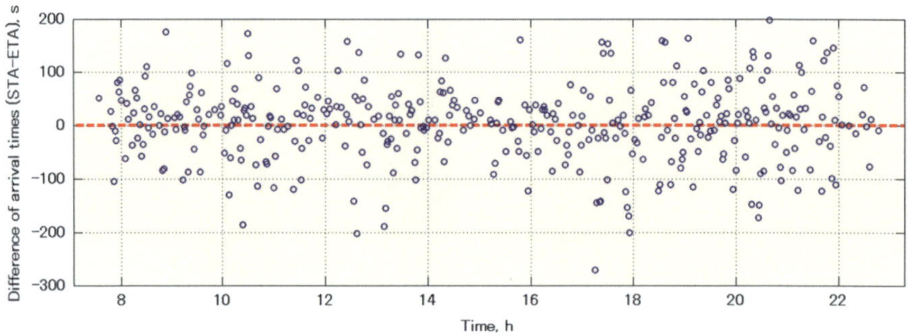

Fig. 17 Difference of arrival times, STA–ETA (ETA, estimated time of arrival)

Figure 16 shows a small part of the time history of the arrival time assignment. The horizontal axis represents the time when the time assignment is performed, and the vertical axis represents the assigned arrival times. The vertical black lines mark the times for which the arrival times were assigned. We can see that the arrival time assignment was performed dynamically. Figure 17 describes the variations of arrival times as the deviations from the horizontal red dashed line. Considering the dispersion of the data away from the red dashed line, it appears that not only the time-delay operation but also the time-advance operation is used to take the advantage of any unoccupied time.

Figures 18, 19, 20, 21, 22, 23, 24, 25, 26, and 27 illustrate the state histories of a particular flight extracted from these flights. The optimal trajectory derived by trajectory optimization shows some important differences from the actual flight trajectory. In the cruise phase, the optimal trajectory chooses a higher altitude than that of the actual flight, obtaining higher flight efficiency by this model. In the

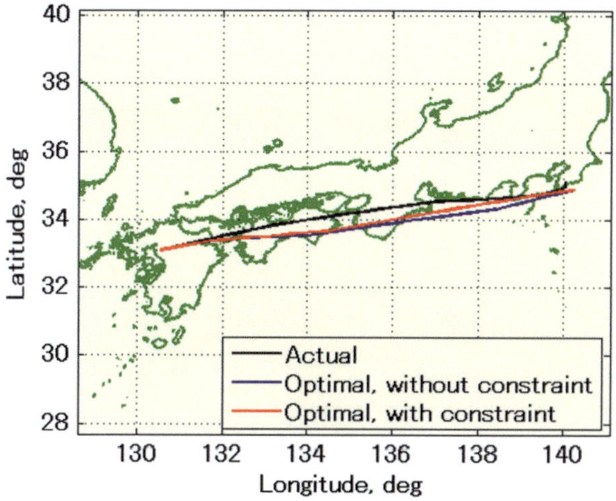

Fig. 18 Flight track

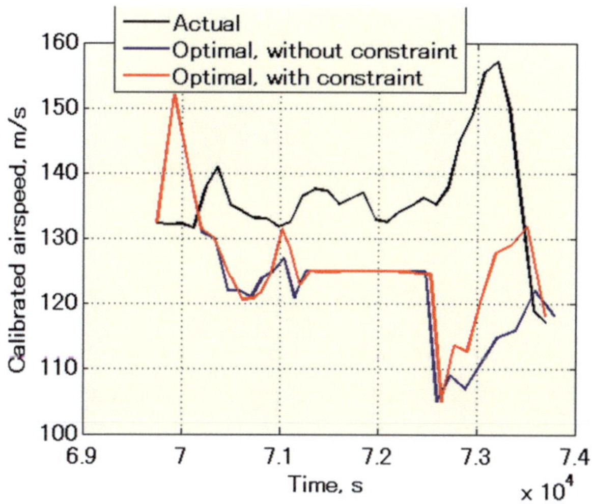

Fig. 19 Calibrated airspeed

descent phase, the optimal trajectory chooses a slower speed than in the actual flight with the reduction of the throttle reducing the fuel consumption. Such flights introduce long descent distance with a shallow flight path angle, leading to an earlier top of descent and a shorter cruise phase where more fuel is consumed thus reducing

Fig. 20 Barometric altitude

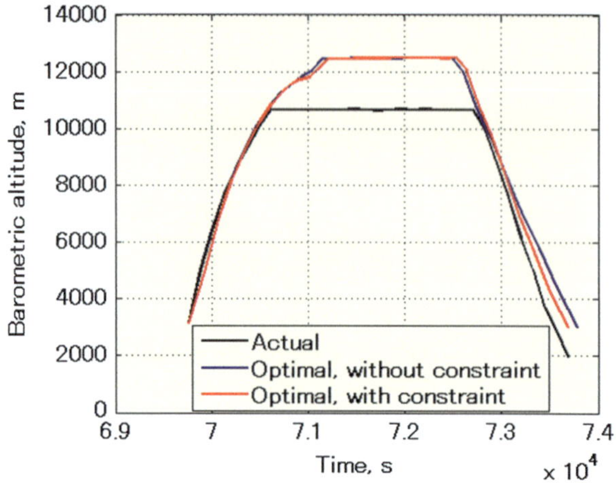

Fig. 21 Flight path angle

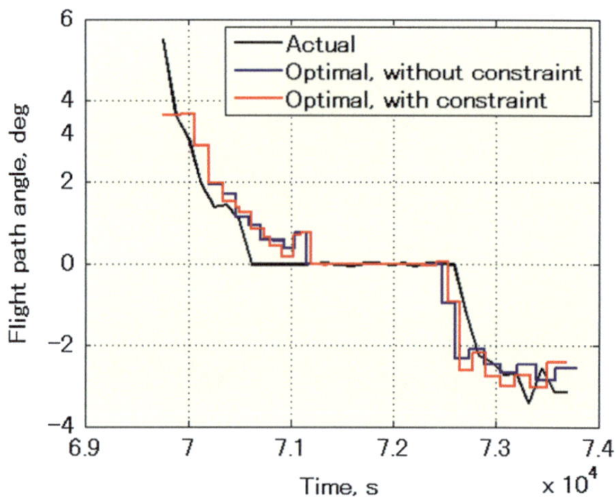

the total fuel consumption. According to the track chart, both of the optimal trajectories choose the great circle course regardless of whether the arrival times are assigned. After receiving the assigned arrival time, the flight arrives at a slightly earlier time than the free optimal flight; arrival time adjustment is achieved not by changing the flight path but mainly by adjusting the speed. The relatively large difference between the two can be recognized in their descent phases; therefore, it appears that the influence on flight efficiency owing to the speed adjustment in the descent phase is smaller than that in the other phases.

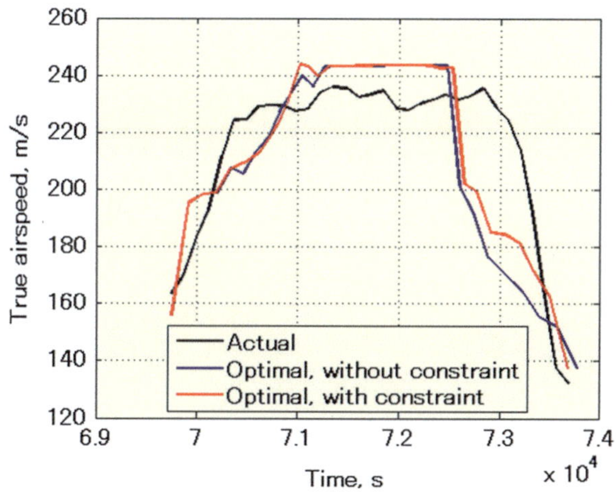

Fig. 22 True airspeed

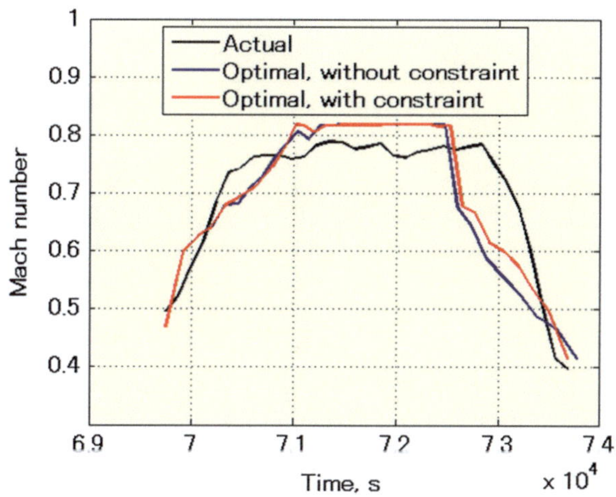

Fig. 23 Mach number

6 Conclusion

In this study, the concept of arrival management that enables arriving aircraft to optimize their flight efficiency was proposed and examined using actual flight track data. This concept is based on the optimization of flight trajectories and arrival time assignment with dynamic programming used for the optimization. At first, each flight trajectory is freely optimized to estimate the potential benefits of the free-flight concept. Next, the constraint of arrival time separation is imposed at a merging point to ensure safe separations among the aircraft in the terminal area. Consequently, optimal trajectories free from conflicts in the terminal area are generated by maximizing the benefits for the flight efficiency criteria.

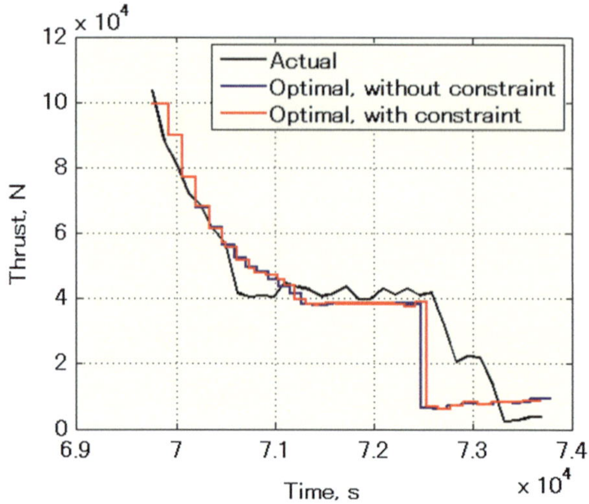

Fig. 24 Thrust

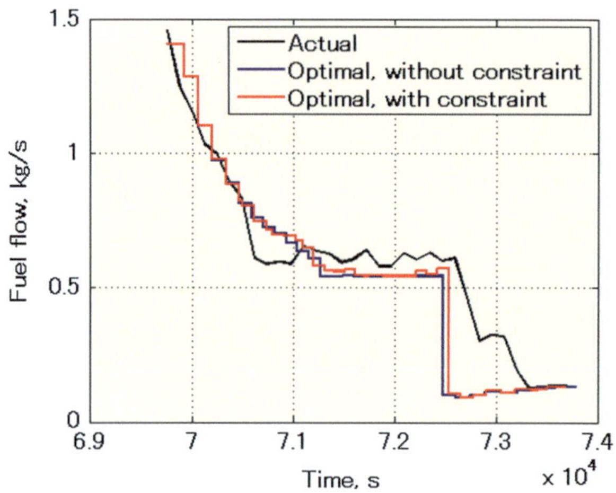

Fig. 25 Fuel flow

We applied this concept to Tokyo International Airport, the most congested airport in Japan, and analyzed the potential benefits of arrival time assignment obtained by the proposed method. The results show that optimal flights with arrival time assignment can produce potential benefits as large as those without the constraint, i.e., arrival time assignment does not significantly reduce the efficiency of the original optimal flight. Furthermore, it became clear that arrival time adjustment is mainly achieved by controlling the speed in the descent phase as it does not have significant impact on fuel consumption.

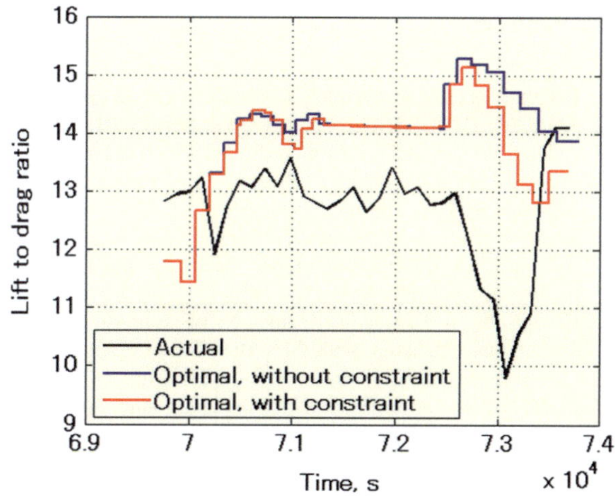

Fig. 26 Lift-to-drag ratio

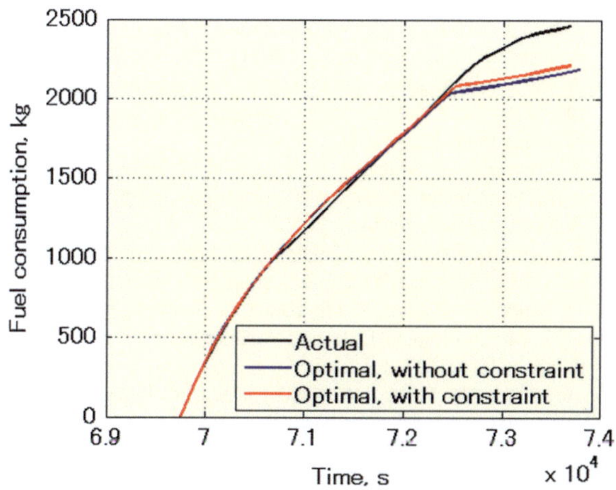

Fig. 27 Fuel consumption

Acknowledgments This research is financially supported by the program of Ministry of Land, Infrastructure, Transport and Tourism in Japan for promoting technological development of transportation. The numerical weather prediction GPV data released by the Japan Meteorological Agency and handled by the Research Institute for Sustainable Humanosphere, Kyoto University and the BADA model developed by EUROCONTROL are effectively used to reconstruct flight parameters from surveillance information data and to optimize flight trajectories. These organizations' support to the research is greatly appreciated.

References

1. JAPAN AIRLINES, "JAL REPORT 2015 (Annual report 2015)," 2015.
2. ANA HOLDINGS INC., "Annual report 2014," 2014.
3. Federal Aviation Administration (FAA), "NextGen Implementation Plan," June, 2013.
4. SESAR Joint Undertaking, "Roadmap for Sustainable Air Traffic Management, European ATM Master Plan, edition 2," October 2012.
5. Ministry of Land, Infrastructure, Transport and Tourism (MLIT), Study Group for the Future Air Traffic Systems, "Long-term Vision for the Future Air Traffic Systems–Changes to Intelligent Air Traffic Systems," 2010.
6. H. Matsuda, A. Harada, T. Kozuka, N. Wickramasinghe, Y. Miyazawa, "Analysis on Airliner Operational Efficiency with GPS Data Based Flight State Estimation and Flight Trajectory Optimization," JSASS Western Branch Conference 2014, Fukuoka, Japan, 2014.
7. Y. Miyamoto, A. Harada, N. Wickramasinghe, Y. Miyazawa, K. Funabiki, "Evaluation Analysis on Airliner Operational Performance with Flight Trajectory Optimization using BADA Aircraft Performance Model," JSASS Aerospace Technology Japan, vol.13, 2014, p.1–10.
8. Y. Miyamoto, N. Wickramasinghe, A. Harada, Y. Miyazawa, K. Funabiki, "Analysis of Fuel-Efficient Airliner Flight via Dynamic Programming Trajectory Optimization," Trans. JSASS Aerospace Technology Japan, vol.11, 2013, p.93–98.
9. A. Harada, T. Kozuka., Y. Miyazawa, N. Wickramasinghe, M. Brown, and Y. Fukuda, "Analysis of Air Traffic Efficiency using Dynamic Programming Trajectory Optimization," ICAS2014, St. Petersburg, Russia, 2014.
10. Y. Miyazawa, H. Matsuda, S. Shigetomi, A. Harada, T. Kozuka, "Potential Benefits of Arrival Time Assignment," Eleventh USA/Europe Air Traffic Management Research and Development Seminar, Lisbon, Portugal, 2015.
11. Eurocontrol Experiment Center, "User Manual for the Base of Aircraft Data (BADA) Revision 3.11, EEC Technical/Scientific Report No.13/04/16-01," 2013.
12. H. Totoki, T. Kozuka, Y. Miyazawa, K. Funabiki, "Comparison of JMA Numerical Prediction GPV Meteorological Data and Airliner Flight Data," Trans. JSASS Aerospace Technology Japan, vol.12, 2013, p.57–63.
13. A. Harada, Y. Miyamoto, Y. Miyazawa, K. Funabiki, "Accuracy Evaluation of an Aircraft Performance Model with Airliner Flight Data," Trans. JSASS Aerospace Technology Japan, vol.11, 2013, p.79–85.

A Dynamic Multi-Commodity Flow Optimization Algorithm for Estimating Airport Network Capacity

Murad Hossain, Sameer Alam, and Hussein Abbass

Abstract Estimating the capacity of an airport network system is an NP-hard problem. It is defined as the maximum traffic that can be accommodated by a network of airports subjected to resource constraints, such as fleet mix and node/link capacity. Mathematically, the problem is modeled as a classical multi-commodity flow (MCF) problem. In MCF it is generally considered that the resources required by the commodities at a node or link cannot change over time and must be independent of the interaction among the commodities. However, in an airport network, the local resource requirements for aircrafts usually change over time due to different weather condition, runway configurations, and different aircraft mix. In addition, in a given airport network, the flow requires a certain amount of time to travel through each link and can't be assumed to travel instantaneously through the network as in the case of an electricity network. These complexities deem existing MCF algorithms inapplicable to estimate the flow capacity of an airport network. To address this problem, we propose a new method to estimate the capacity of an airport network and develop a dynamic multi-commodity flow optimization algorithm. The proposed optimization algorithm is augmented by an iterative Hill-Climber algorithm to solve the network capacity model in which all flow constraints of air traffic are preserved. Experimental results show that the proposed model is not only capable of realistically estimating the airport network capacity under different levels of aircraft mix but also in identifying individual flows at different links and amount of delay for each aircraft.

Keywords Network capacity • Airport network • Multi-commodity • Flow optimization

---

M. Hossain (✉) • S. Alam • H. Abbass
School of Engineering and Information Technology, University of New South Wales (Canberra Campus), Campbell, NSW, Australia
e-mail: m.hossain@adfa.edu.au; s.alam@adfa.edu.au; h.abbass@adfa.edu.au

Electronic Navigation Research Institute, *Air Traffic Management and Systems II*, Lecture Notes in Electrical Engineering 420, DOI 10.1007/978-4-431-56423-2_11

1 Introduction

The capacity of flow networks indicates the maximum attainable throughput of a network under normal operating conditions (i.e., without jamming or congestion). A good network capacity estimation model would enable us to predict how much additional demand can be accommodated by a network and hence establish an efficient policy for traffic restraint and growth. Furthermore, it can enable us to determine what educative steps should be taken to prepare for the time when additional capacity will be required to accommodate future growth. Network capacity estimation is an NP-hard problem. Pioneering work to solve the network capacity problem traces back to Ford and Fulkerson [10], who developed a labeling algorithm for the network maximum-flow problem on the basis of max-flow min-cut theory. However, this algorithm is ideal to solve problems with a single origin-destination [15]. In 1972 Iida [12] developed an incremental assignment approach in which a certain portion of origin-destination (OD) demand matrix was iteratively added to the network. On the basis of the updated link cost (travel times), a link was eliminated from the network if it reached its capacity threshold. Finally, the network capacity was obtained when a certain OD pair was not connected anymore. The main drawback of this approach is the choice of the route. Realizing the effects of route choice behavior and congestion, Asahura and Kashiwadani [1, 2] proposed a bi-level programming approach in which routing strategies and congestion effects were explicitly considered. In a recent study, Chen et al. [4] developed a bi-level model to deal with the network turning restriction design problem in which entities are prohibited to drive into restricted downstream links at a group of intersections. Along with the bi-level model, the importance of developing probabilistic procedures for the quantitative evaluation of capacity [6, 7] and flexibility [4] of transportation networks capacity has also been investigated in recent years.

In this paper, we define an airport network as a system of airports interconnected by air routes. Despite the significant contribution of previous studies to various aspects of network capacity modeling, a comprehensive approach to investigate the flow capacity of an airport network is lacking. A few number of attempts have been made to estimate the total capacity of an entire airport network system for any region or country [8, 9]. There is a growing concern among airlines and small group of policy analysts that the airport network is running out of capacity [8, 14].

Conventionally in traffic flow networks, the capacity is estimated using a multi-commodity flow (MCF) model [5]. This method is not directly applicable to airport networks for the following reasons: (a) movements in an airport network involve flows of aircrafts with different speeds; (b) flow is heterogeneous given different wake vortices categories of aircraft, viz., light, medium, and heavy; (c) different types of aircrafts require different amount of resources at landing and departure airports; (d) there must be a minimum separation distance between two consecutive aircrafts, which depends on the type of operation (landing or takeoff) and the preceding aircraft type and operation for managing wake vortices; (e) aircrafts departing from an airport are expected to land at destination airports within a time window, because aircrafts cannot hold in air indefinitely; and (f) multiple origin-destination (OD) pairs exist and flows between different OD pairs are not exchangeable or substitutable in an airport network capacity problem. These characteristics make airport network capacity modeling a complex, yet interesting, problem to solve.

This paper attempts to solve this problem and proposes a model and an optimization algorithm to estimate the maximum capacity of an airport network. We borrow the classical MCF problem concept to formulate the problem, where we represent flows between two nodes (airports) as different commodities and the problem is formulated using a time slot of 1 h. This hourly rate of flow (landings and takeoffs) is bounded by a capacity constraint. The proposed model also considers wake vortex interactions among aircrafts of different categories (heavy, medium, light) during landing and takeoff at a given airport. The underlying premise is that the flow in an airport network can be modeled as a multi-path, steady state network of queues, whose maximum capacity is the sum of maximum airport operational rates.

This paper is organized as follows. Section 2 describes the problem formulation. Section 3 explains the steps involved in solving the capacity estimation problem. Section 4 presents the experimental setup. Section 5 presents the flow characteristics and the results obtained. Section 6 draws conclusions.

2 Problem Formulation

A graph $G(V, E)$ is used to describe an airport network in which a node set (V) and an edge set (E) represent n airports and m direct flight routes between these airports, respectively [11]. The network is encoded using an adjacency matrix ($A_{n \times n}$) such that $a_{ij} = 1$ if a flight link exists between airport-pair i and j, otherwise $a_{ij} = 0$. In such network, an OD pair is defined as those nodes that are directly connected by an edge. For each of the OD pairs, traffic in the form of aircraft is entering into the network through origin nodes and travel along links to destinations. A commodity in an airport network is the flow from a given node to one of its neighboring node. Since in a network there are $m = |E|$ directly connected OD pairs. Thus the total number of commodities will be $K = 2 \times m$. The question arises as to how we can define/determine an individual airport capacity.

Under normal conditions, an airport capacity heavily depends on the traffic mix. The traffic mix, consisting of the aircraft possessing different weights and speeds, requires different rules (separation) to maintain safe time/distance interval between subsequent aircrafts during landing and taking off. In this study, we have used a time-based separation minimum to avoid the wake-vortex turbulence which is given in Table 1.

Two runways with identical operating conditions may result in different capacities/throughputs due to aircraft sequencing difference. Thus, we have considered the local capacity of an airport as the total available slots for landing and takeoff. In our

Table 1 Separation minima (in minutes) between aircrafts considered in this paper

Separation minima (arrival-departure)			
Leading aircraft	Trailing aircraft		
	L	M	H
L	2	2	2
M	2	2	2
H	2	2	3
Separation minima (departure-arrival)			
Leading aircraft	Trailing aircraft		
	L	M	H
L	2	2	2
M	2	2	2
H	2	2	3
Separation minima (arrival-arrival or departure-departure)			
Leading aircraft	Trailing aircraft		
	L	M	H
L	2	2	2
M	3	2	3
H	3	2	3

formulation, we have defined airport capacity as the total number of slots, where each slot has equal time duration. In other words, we define a node capacity as the total available time. Since our intention is to measure the hourly flow capacity of an airport network, we define the node capacity $C(i) = 60$ slots for all nodes in the network where each slot is a one-minute duration.

Notations The formulation of an airport network capacity problem requires definition of the following notations:

$t =$ a positive integer which represent the hour of operation
$K =$ the set of commodity
$IK =$ the set of initialized commodity
$UK =$ the set of un-initialized commodity

For each $k \in K$

$s^k =$ source or origin node of commodity k
$d^k =$ destination node of commodity k
$F_i^{k+}(t) =$ inflow of commodity k of node i during operation hour t, which is a set of arrival flights
$F_i^{k-}(t) =$ outflow of commodity k of node i during operation hour t, which is a set of departure flights
$F_i^{+}(t) = \sum_k F_i^{k+}(t)$, total inflow of node i for operation hour t
$F_i^{-}(t) = \sum_k F_i^{k-}(t)$, total outflow of node i for operation hour t
$u_i(k)$ represents the resource used by commodity k at node i

Let f denote a flight that is scheduled to operate during a day.

Then

$D(f)$ = destination airport of f
$O(f)$ = origin airport of f
$TD(f)$ = schedule time of departure of f from $O(f)$
$TA(f)$ = schedule time of arrival of f at $D(f)$
$wv_f \in \{L, M, H\}$ = wake vortex class of f, where L, M, and H represent the light, medium, and heavy aircraft, respectively
$Op(f) \in \{A, D\}$ = arrival (landing) and departure (takeoff) operation of f
$S(a, b, c, d)$ = separation distance in time between the leading and trailing aircraft, where a and b represent the type of leading and trailing aircraft and c and d represent takeoff or landing of respective flight
$delay(f)$ = delay of f at $D(f)$
t_{ij} = required travel time for a flight from airport i to j

Let $\phi\left(F_i^+(t), F_i^-(t), S\right)$ be a slot-assignment function that returns the minimum time required to accommodate the inflow $F_i^+(t)$ and outflow $F_i^-(t)$ at a node i during an hourly interval t given the separation matrix S. The separation matrix S has four dimensions: leading aircraft type, trailing aircraft type, leading aircraft operation, and trailing aircraft operation. For example, $S[L][M][A][D]$ defines the separation distance in minutes between a light aircraft landing at a node followed by a departure of medium aircraft. With the help of the function $\phi()$ and separation matrix S, we can define the departure-arrival constrains as $\phi\left(F_i^+(t), F_i^-(t), S\right) \leq C(i), \forall i$. In an airport network, an aircraft departing from an airport is expected to land at a destination airport within a time frame. This travel time introduces a time dimension into the problem formulation. We call this constraint as the timing constraint.

Let t_{ij} denote the required travel time for a flight from airport i to j. We assume that the travel time for all type of aircrafts are equal. Let $TD(f)$ and $TA(f)$ denote the time of departure (takeoff) and time of arrival (landing) of a flight $f \in F_i^{k-}(t)$ from node i to j, where $k \in K$ denotes the commodity from node i to j and the minus symbol "–" means an outflow from node i. Then the timing constraint is defined as $TA(f) = TD(f) + t_{ij}, \forall f \in F_i^{k-}(t)$. This timing constraint makes the problem very hard to solve. In a real traffic scenario, there is a common practice to associate some delay to a flight, landing at the destination airport, for capacity improvement and maintaining separation safety. We can modify the timing constraint as $TA(f) = TD(f) + t_{ij} + delay(f), \forall f \in F_i^{k-}(t)$, where $delay(f)$ represents the delay of flight f and its value is bounded by $0 \leq delay(f) \leq 15$ min. The delay variable $delay(f)$ brings some flexibility in the timing constraints and helps to find better feasible solutions for an optimization method. The airport network capacity model can be formulated as follows:

$$maximize : \sum_k \sum_i F_i^{k-}(t) \tag{1}$$

Subject to

$$TA(f) = TD(f) + t_{ij} + delay(f), \quad \forall (i, j, k) \tag{2}$$

$$\phi\left(F_i^+(t), F_i^-(t), S\right) \leq C(i), \quad \forall i \tag{3}$$

$$F_i^{k+}(t) \geq 0 \forall (i, k) \tag{4}$$

$$F_i^{k-}(t) \geq 0 \forall (i, k) \tag{5}$$

$$\sum_t \sum_k \sum_i F_i^{k-}(t) = \sum_t \sum_k \sum_i F_i^{k+}(t), \quad \forall (i, k, t) \tag{6}$$

where $TD(f), TA(f)$ denote the departure and arrival time of a flight $f \in F_i^{k-}(t)$ from node i to j, $F_i^+(t) = \sum_k F_i^{k+}(t), \quad \forall i$, and $F_i^-(t) = \sum_k F_i^{k-}(t), \quad \forall i$.

3 Heuristic Solution Approach

A heuristic approach is developed in which a certain amount of flow is incrementally added to an initial feasible solution until the network reaches its capacity. The heuristic solution approach consists of two key modules: (i) an initial feasible solution generator function and (ii) an iterative solution improvement method to increase the traffic flow in the network subject to node capacity and departure-arrival timing constraints.

3.1 Initial Feasible Solution Generation

In an airport network capacity estimation problem, the departure-arrival timing constraint Eq. (2) (all departure flights from a node must have to land at the destination at a specific time within a maximum delay of 15 min) and the separation minima between two aircrafts at a node make a random initial solution very hard to be feasible. In order to generate an initial feasible solution, we divide a 1 h time window into 60 equal slots. We call these 60 slots together as "slot box" and represent it as SB. Each SB has a unique identification number for each airport to represent an hour of operation. For example, an *SB(i,5)* represents the flow (both departure and arrival) of node *i* during the operational hour of 5:00 am. To simulate a full-day operation in a network, we need 24 SBs for each node.

The flow of a commodity consists of light, medium, and heavy aircrafts being placed in the slots of the source and destination node such that the separation between two consecutive flights is maintained and all departed aircrafts can land at the corresponding destination within a maximum delay of 15 min. Once an aircraft is inserted into an SB of the departure and that of landing node, it is copied to all succeeding slot boxes of those nodes. For example, if a light aircraft is placed at the fifth slot of SB(i,1) for departure and fifth slot of SB(j,2) as an arrival flight, then all operating hours $t \geq 1$ at node i must have a departure flight at its fifth slot and at node j for all operating hours $t \geq 2$ also have a arrival at fifth slot. This means that we consider a continuous deterministic flow, which is a common practice to determine the capacity upper bound of a network [13, 15].

We assume that resources of a node are equally shared by its commodities. That is, the hourly resource of 60 slots of a node is equally shared by the commodities associate with its links. If $r_i(k)$ represents the amount of shared of node i assign to a commodity k, then the following condition hold for all nodes:

$$C(i) = \sum_k r_i(k), \forall i \tag{7}$$

The initial feasible solution generation algorithm is presented as a flowchart in Fig. 1, where $u_i(k)$ represents the resource used by commodity k at node i and IK, UK represent the set of initialized and un-initialized commodity.

3.2 *Iterative Solution Improvement*

After generating an initial feasible solution, the network capacity can be obtained using an iterative process, which consists of the following modules, which were designed to increase the number of hourly aircraft movements.

3.2.1 Shifting

Shifting is an operation to move a flight from its current slot to an earlier slot such that: (i) the separation minima between its adjacent flights are maintained and (ii) the delay of the flight remains within the bound 0–15 min. Figure 2 shows an example of a shifting operation of a slot box. In the illustration of a shifting operation in Fig. 2, the flight $f2$ is shifted to slot 3, which decreases its delay to 9 min, and the separation distance between f1 and f2 satisfies the minimum required distance of 2 min for a light-light departure-arrival pair, whereas flight f3 which is departing from slot 9 cannot shift to an earlier slot because an increase in its delay will occur beyond the maximum limit of 15 min.

3.2.2 Swapping

Swapping is an operation to change sequence of the flights in an SB. Swapping is considered as a hill-climbing operation to improve the quality of the solution, which basically changes the slot of two flights at a time and continues the process until no

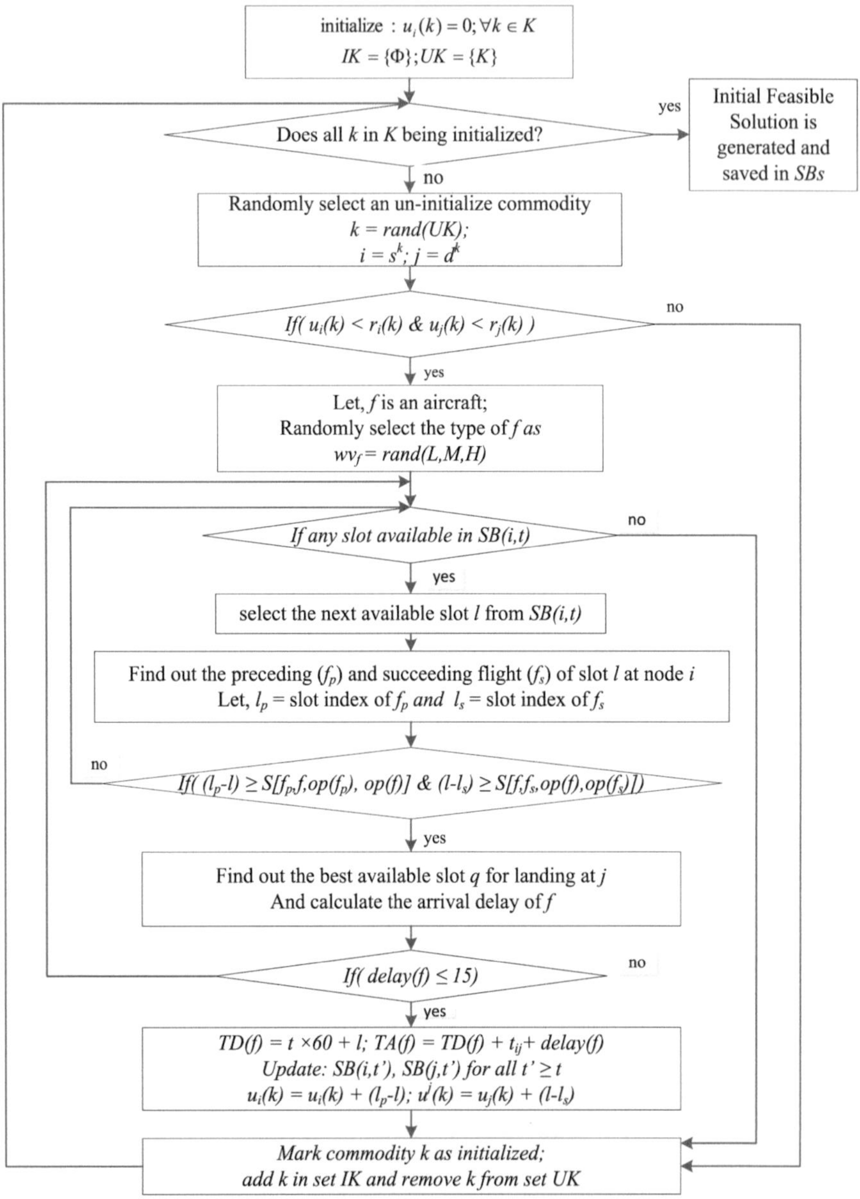

Fig. 1 Initial feasible solution generation process

further improvement is possible while maintaining separation minima and the travel time constraint.

Figure 3 shows an example of swapping operation between flight f1 and f3. The purpose of a swapping operation is to make some free slot(s) so that one or more

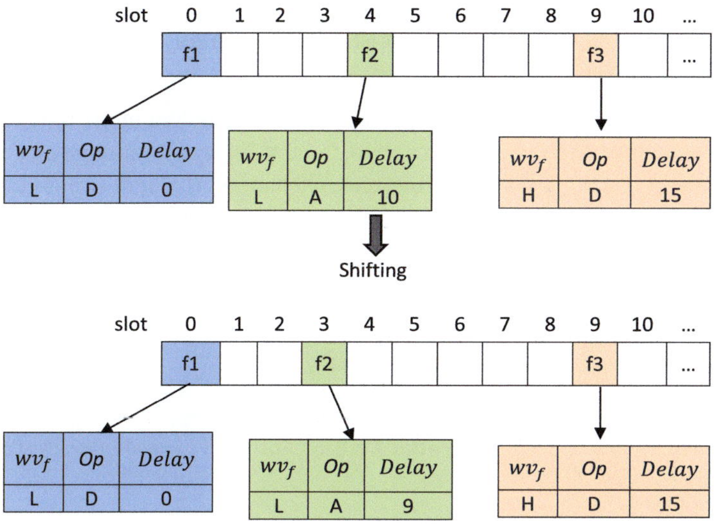

Fig. 2 An illustration of shifting operation in an SB

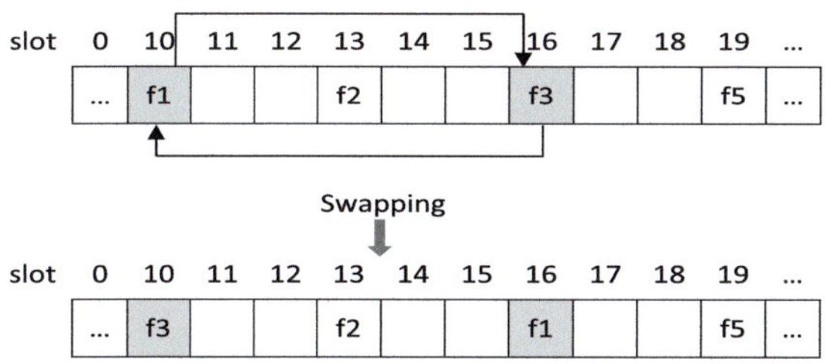

Fig. 3 An example of swapping operation

flights can be shifted to an earlier slot. In some cases, a swap can increase delay. We maintain all constraints satisfied and only allow swapping to occur between two flights if and only if the new sequence does not violate any constraint.

3.2.3 Inserting

In an initial feasible solution, the operations of swapping and shifting may create some free slot where we can insert some new flight(s). A new flight is inserted into SBs of source and destination nodes that fulfil the following conditions:

(a) There is enough room between two flights, at the source node, at the end, or at the beginning of the slot such that the aircraft maintains separation minima with adjacent flights.

(b) There is also a slot q that is available for landing while meeting the separation minima and satisfying the timing constraint Eq. (2) for any of the neighboring flights ($j \in Ng(i)$) of the source.

The type of the newly inserted aircraft is selected randomly. If the new flight cannot be added to the network due to violation of separation minima constraint or its type, then an attempt is made to insert a light aircraft. If a light aircraft cannot be added at a node i as a source, then it confirms that there is no room at the slot box of node i to accommodate extra flow and the operation gets abundant.

Algorithm 1 Solution improvement (FS)

procedure SOLUTION IMPROVEMENT($FS, G(V, E), t$)
    FS is an initial feasible solution for the network capacity problem
    $G(V, E)$ is airport network having n nodes and K commodities
    t is the earliest operation hours where the flow of in $G(V, E)$ reached a steady state.
    $Ng(i)$ neighbor of node i
    $isImprove = true$
    while $isImprove$ **do**
        $isImprove = false$
        $i = 1$
        while $i < n$ **do**
            Repeat for all
            $f_a \in SB(i, t) \& f_b \in SB(i, t) | f_a \neq f_b$,
            if swap(f_a,f_b) in $SB(i, t)$ **then**
                perform shifting at node i
                for all $j \in Ng(i)$ **do**
                    if insert(f_{new}, $i \rightarrow j$) of $SB(i, t)$ **then**
                        $isImprove = true$
                    end ifend if
                end for
            end ifend if
            end loop
            $i = i + 1$
        end whileend while
    end whileend while
    return *solution*
end procedure

After applying the swapping, shifting, and inserting operations when no further improvement is possible for a given sequence (solution), the value of the function $\phi()$ Eq. (3) can be determined by simply calculating the time of the operation of the last flight in the sequence at a given node. Based on swapping, shifting, and inserting operations, the quality of an initial feasible solution can be improved. Algorithm 1 illustrates the procedure to improve a feasible initial solution for the network capacity estimation problem.

Once the initial feasible solution gets improved by the solution improvement Algorithm 1, the hourly capacity of the network gets calculated by counting the total number of flight movements in the solution, which provides a tight capacity upper bound of the steady-state flow.

4 Experimental Setup

To illustrate the effectiveness and applicability of the proposed method, numerical studies on two different networks were done. In these test networks, all the nodes have a single runway. In our experiments, the operating condition of the nodes does not change over time, i.e., the separation minima remain unchanged. Each evaluation is repeated 30 times with different seeds.

4.1 Test Networks

In order to assess the effectiveness of the proposed airport network capacity estimation model, we perform experiments on two different types of network. First, we applied the proposed procedure to a simple network shown in Fig. 4. Our first experimental network consists of three nodes and six directed links, which is shown in Fig. 4. We named this network as "network-I." Network-I is a weighted directed graph. The weight of a link represents the travel time between its start and end node.

Apart from the simple network shown in Fig. 4, we also test the proposed model with a much complicated real-world network. This real-world network is based on the Australian airport network which we called "network-II." The Australian airport network is a very large network and has many peripheral airports that carried out only a small amount of flights. These peripheral airports make limited contributions to the overall network capacity. The capacity bottleneck mainly depends on hub airports. More details on the characteristics of the Australian airport network can be found in [11]. Figure 5 shows network-II, representing those Australian airports that operate, on average, more than five flights daily. In Fig. 5, the size of a node is proportional to the number of its direct connections with other nodes in the network, which is called the degree [3]. In this network, nodes 8 and 11 have a degree of 20 (10 in-degree and 10 out-degree) which is the highest in the network, whereas node 3

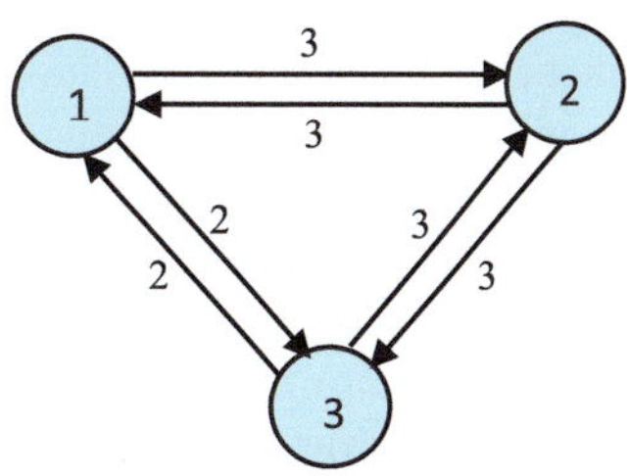

Fig. 4 Example network-I, a fully connected network of three identical nodes

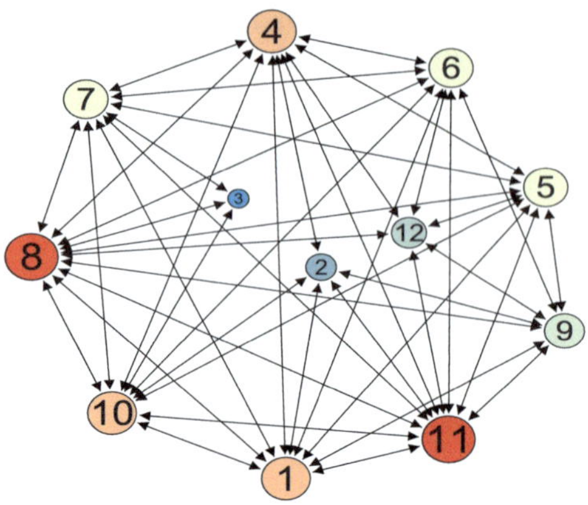

Fig. 5 Example network-II, network of hub nodes of Australian airport network

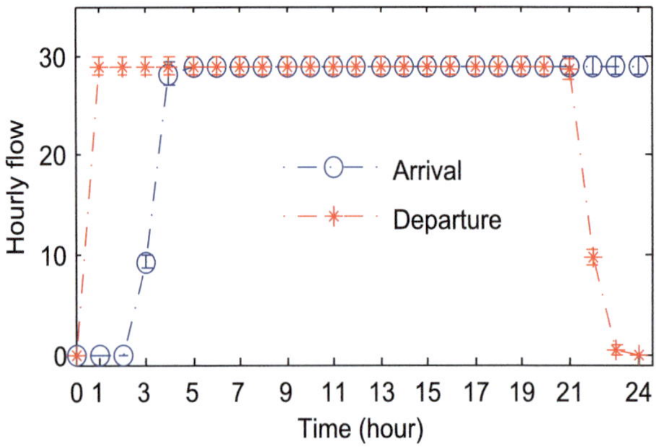

Fig. 6 Hourly traffic flow in network-I over a period of 24 h

has the lowest degree of 6. In this network the travel time of the links is set randomly between 1 and 4 h.

5 Results and Analysis

We first present hourly flows in network-I over a period of 24 h time. As the objective of capacity estimation is to estimate the maximum attainable flow in an airport network, we attempt to detect the departure sequence at the nodes that remain unchanged and achieved maximum steady-state flow. Figure 6 shows the hourly departure and arrival in network-I. At early hours, there is no arrival in the network. As time goes, the departed flights in early hours start to arrive at destinations. The load on the network continues to increase until a steady state is reached. If one can find the right departure sequence, then it is possible to get a maximum steady-state flow representing the total capacity of the network. Flow in network-I reaches a steady state from the fifth hour of operation.

In any airport network, the total flow consists of a number of light, medium, and heavy aircrafts. Thus, to better investigate the capacity of a network, we calculate the actual number of light, medium, and heavy aircrafts. Table 2 summarizes the capacity of the example networks, and Figs. 7 and 8 show the total flight movements throughout a day.

To get an insight of the efficiency of the proposed model, we measure the number of unused slots for every node in the network. The number of unused slots is calculated from the flight sequence in SBs. It is the difference between the slot positions of consecutive flights and the minimum separation required between them. For example, if two light aircrafts are departing from slots 5 and 9, respectively, then the distance used by these two flights is four slots where the minimum separation required between them is two slots (since each slot is equivalent of 1 min). So the

Table 2 Hourly capacity of test networks (average of 30 different runs)

Network	Capacity	#L	#M	#H
I	57.7 ± 1.34	38.7 ± 7.60	10.6 ± 4.01	8.4 ± 3.87
II	222.2 ± 21.04	111.7 ± 33.01	57.5 ± 20.22	53.0 ± 20.21

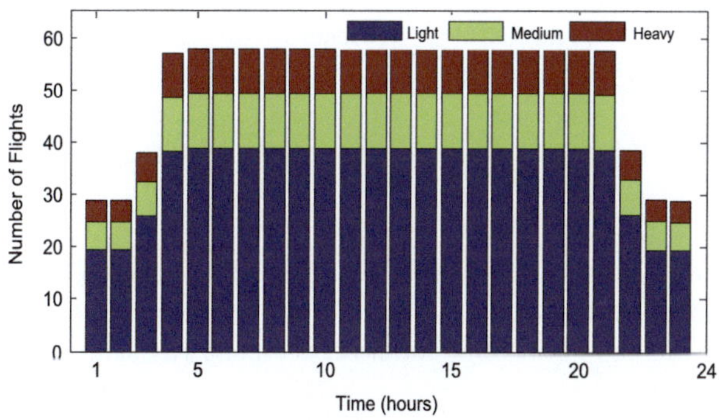

Fig. 7 Hourly traffic of network-I

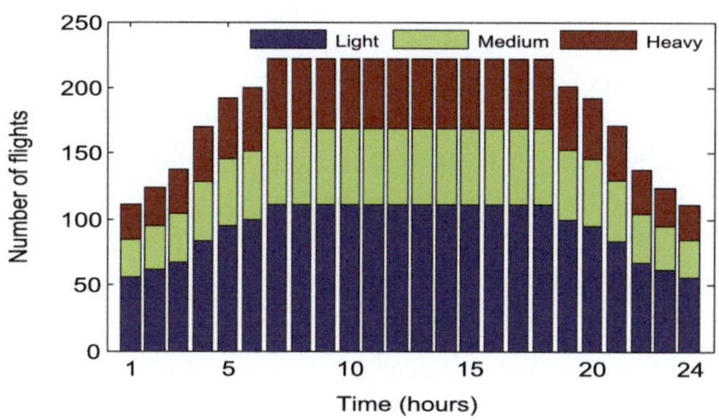

Fig. 8 Hourly traffic of network-II

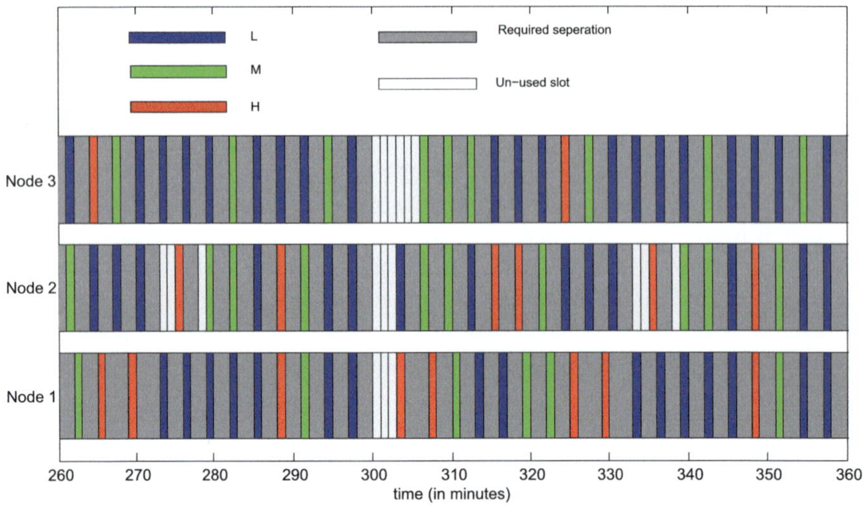

Fig. 9 Status of the nodes in network-I during steady state

Table 3 Summary of the node's hourly utilization of network-I at steady state

Node	Total flow	Lihgt (L)	Medium (M)	Heavy (H)	Unused slots
1	19.35 ± 0.99	12.90 ± 2.89	3.50 ± 2.01	2.95 ± 1.19	2.35 ± 1.76
2	19.35 ± 0.87	12.60 ± 3.25	3.85 ± 1.53	2.90 ± 1.94	2.00 ± 1.65
3	19.00 ± 0.56	13.20 ± 2.48	3.25 ± 1.45	2.55 ± 1.54	2.65 ± 1.50

number of unused slots in this case is two (four minus two). Figure 9 shows the status of the nodes in network-I during a portion of steady state. From Fig. 9 it is noticeable that there is not enough room between any two nodes to accommodate more flights.

Tables 3 and 4 summarize the hourly uses of the nodes of network-I and network-II, respectively, during a steady-state network operation. From Tables 3 and 4, we can see that the airport's (nodes) slots are fully utilized. Node 3 in network-II maintains some unused slots because of the very limited number of connections it has compared to other nodes in the network.

We also analyzed the total delays associated with the estimated capacity in Table 5. The average delay per flight is found to be around 5 min for both networks. This is well accepted in a usual air transportation system. Among the total flights, only about 20% of them get delayed more than 10 min, whereas most of them have delays less or equal to 10 min.

Table 4 Summary of the node's hourly utilization of network-II at steady state

Node	Total flow	#L	#M	#H	# Un-used slots
1	19.25±2.34	8.70±3.28	5.60±1.82	4.95±1.67	1.15±0.88
2	17.65±1.69	11.35±2.46	3.50±1.67	2.80±1.01	6.55±3.70
3	14.8 ±1.85	8.30±1.90	3.50±1.28	3.00±1.30	14.15±5.88
4	19.4 ±3.20	8.40±3.66	5.50±2.48	5.50±2.46	2.50±1.80
5	18.95±0.69	9.30±2.18	5.45±1.36	4.20±1.91	2.05±1.54
6	18.65±0.99	9.15±2.50	4.80±1.32	4.70±1.75	3.05±2.50
7	18.8 ±0.62	8.00±2.13	6.00±1.69	4.80±1.61	2.15±1.31
8	20.05±3.20	9.20±4.01	5.30±1.69	5.55±2.14	1.85±1.39
9	19.65±2.80	11.80±3.37	3.90±1.53	3.95±1.28	3.05±2.09
10	18.85±1.04	7.90±2.29	5.65±2.23	5.30±1.63	2.65±1.90
11	18.9 ±1.41	8.65±2.60	5.00±1.56	5.25±1.86	2.60±1.64
12	17.25±1.21	10.95±2.63	3.30±1.59	3.00±1.59	7.25±3.54

Table 5 Summary of delay of test networks

Network	Total flight movement	Delay(D)per flight (min)	Number of flights delayed ($0 \le D \le 10$)	Number of flights delayed ($10 < D \le 15$)
I	615.55 ± 13.88	4.72 ± 1.13	271.6 ± 53.67	18.9 ± 56.31
II	2271.15 ± 40.91	5.13 ± 0.44	912.1 ± 124.3	516.8 ± 99.87

6 Conclusion

This paper has presented preliminary evaluation of the proposed airport network capacity model and a heuristic algorithm. One of the limitations of the proposed method is that the nodes in an airport network can only have one runway. More extensive tests or evaluations need to be performed in future work to assess the effectiveness of the proposed model, including (i) investigating more complex network configurations, (ii) multiple runways in the airports, and (iii) including various demand patterns and demand or preference of fleet mix.

Acknowledgements This work was partially supported by research grants ARC LP110100488 and MIDRMA RG151448.

References

1. Y. Akamastu and O. Miyawaki. Maximum network capacity problem under the transportation equilibrium assignment. *Infrastructure Planning Review*, 12:719–729, 1995.
2. Y. Asakura and M. Kashiwadani. Estimation model of maximum road network capacity with parking constraints and its application. *Infrastructure Planning Review*, 11:129–136, 1993.

3. S. Boccaletti, V. Latora, Y. Moreno, M. Chavez, and D.-U. Hwang. Complex networks: Structure and dynamics. *Physics reports*, 424(4):175–308, 2006.
4. A. Chen and P. Kasikitwiwat. Modeling capacity flexibility of transportation networks. *Transportation Research Part A: Policy and Practice*, 45(2):105–117, 2011.
5. A. Chen, H. Yang, H. K. Lo, and W. H. Tang. A capacity related reliability for transportation networks. *Journal of advanced transportation*, 33(2):183–200, 1999.
6. A. Chen, M. Tatineni, D.-H. Lee, and H. Yang. Effect of route choice models on estimating network capacity reliability. *Transportation Research Record: Journal of the Transportation Research Board*, (1733):63–70, 2000.
7. A. Chen, H. Yang, H. K. Lo, and W. H. Tang. Capacity reliability of a road network: an assessment methodology and numerical results. *Transportation Research Part B: Methodological*, 36(3):225–252, 2002.
8. G. L. Donohue. A simplified air transportation system capacity model. *Journal of Air Traffic Control*, 1999.
9. G. L. Donohue. A macroscopic air transportation capacity model: Metrics and delay correlation. In *New Concepts and Methods in Air Traffic Management*, pages 45–62. Springer, 2001.
10. L. R. Ford and D. R. Fulkerson. Maximal flow through a network. *Canadian journal of Mathematics*, 8(3):399–404, 1956.
11. M. Hossain, S. Alam, T. Rees, and H. Abbass. Australian airport network robustness analysis: a complex network approach. In *Australasian Transport Research Forum (ATRF), 36th, 2013, Brisbane, Queensland, Australia*, 2013.
12. Y. Iida. Methodology for maximum capacity of road network. *Transaction of Japan Society of Civil Engineers*, 205:147–150, 1972.
13. Z. Luo, Y. Liu, and J. Yu. Estimation of urban transportation network capacity considering traveler road preferences. *Journal of Urban Planning and Development*, 138(2):133–142, 2011.
14. National Civil Aviation Review Commission (US). *Avoiding aviation gridlock & reducing the accident rate: A consensus for change.* National Civil Aviation Review Commission, 1997.
15. K. A. Ravindra, T. L. Magnanti, and J. B. Orlin. *Network flows: Theory, algorithms, and applications.* Prentice Hall Englewood Cliffs, 1993.

Part IV
Enablers

Downlink Aircraft Parameter-Based High-Accuracy Tracking System for Air Traffic Surveillance

Xiaodong Lu, Tadashi Koga, and Yoshio Kosuge

Abstract With the rapid increase in air traffic demands, the more accurate and reliable tracking systems for aircraft surveillance are required to improve the capacity, safety, and efficiency of air traffic control (ATC) services. As a suboptimal hybrid filter, the Interacting Multiple Model (IMM) estimator has been applied in the practical aircraft tracking systems. However, it is difficult for the standard IMM filter to precisely estimate the aircraft state when the target is maneuvering since the detection of maneuvers is often delayed by the response of Kalman filters. On the other hand, in the Mode S radar Enhanced Surveillance (EHS), the downlink aircraft parameters (DAPs) are available for obtaining the updated states of aircraft; however, this information of aircraft parameters has not been used systematically to improve the performance of current tracking systems. In this paper, a DAPs-based tracking system which is able to dynamically revise the mode probabilities of the IMM estimator according to the real-time change in the aircraft motion is proposed. The results of computer simulations and practical experiments show the effectiveness of the proposed system by comparing it with the standard IMM-based tracking system.

Keywords Tracking system • Mode S radar • DAPs • IMM

1 Introduction

In response to the rapid increase in air traffic demands, a high-accuracy tracking system, which can achieve accurate and continuous aircraft surveillance in the dynamic changing environment, is required to improve the capacity, safety, and efficiency of air traffic control (ATC) services. With the advancement of the sensor and communication technologies, the extensive deployment of air-to-ground data

X. Lu (✉) • T. Koga • Y. Kosuge
Surveillance and Communications Department, ENRI, 7-42-23 Jindaiji-Higashimachi, Chofu, Tokyo, 182-0012, Japan
e-mail: luxd@mpat.go.jp

Electronic Navigation Research Institute, *Air Traffic Management and Systems II*, Lecture Notes in Electrical Engineering 420, DOI 10.1007/978-4-431-56423-2_12

links leads to the emergence of complementary methods for the aircraft tracking system.

As a considered part of the basis of the surveillance infrastructure, Secondary Surveillance Radar (SSR) Mode S has been widely utilized. There are two possible configurations for the Mode S radar: the first one is called the Mode S Elementary Surveillance (ELS) and the other one is the Mode S Enhanced Surveillance (EHS) [1]. Each aircraft is assigned one unique ICAO 24-bit address by the state registration authority [2]. This address is the one used for selective interrogation, which permits the obtaining of the flight ID and the barometric altitude in 25 ft steps. The Mode S EHS consists of ELS supplemented by the extraction of downlink aircraft parameters (DAPs) which can be used to improve the performance of current tracking systems. Most Mode S radars are applied to aircraft surveillance in terminal and en route areas. Therefore, the DAPs-based tracking system can be applied in these areas, especially in the ascent and descent en route segments.

In [3], the roll angle is used to detect the maneuvering of aircraft and calculate the lateral acceleration to improve accuracies by switching from a straight-line first-order Kalman filter processing positional measurements only to the second-order extended Kalman filter that operates on the same axes and estimates the same quantities plus lateral acceleration. However, by using the roll angle, except turning models, it is difficult to detect other maneuvering motions of aircraft in the real applications. And because the roll angle is easily affected by the outside factors, it is difficult to assure the quality of roll angle measurements. Moreover, compared with the acceleration model, the coordinated turn model can get more improvement for turning motion [4]. And in [5], the DAPs of ground speed, track angle rate, and true track angle are applied to calculate the control input of the Kalman filter, but the improvement is limited by using a single model of the Kalman filter.

The Interacting Multiple Model (IMM) estimator is a suboptimal hybrid filter that was shown to achieve an excellent compromise between performance and complexity [6, 7] and has been applied in the practical aircraft tracking systems [8]. However, it is difficult to accurately estimate the aircraft state when the target is maneuvering since the detection of maneuvers is often delayed by the response of Kalman filters. In this paper, to improve the accuracy, the DAPs-based IMM tracking system is proposed. The system consists of an IMM filter with three different models and a maneuver detector which is able to dynamically revise the mode probabilities in real time according to the different motions of aircraft by using DAPs data. Moreover, as the availability and certification of DAPs cannot be guaranteed, the system with two tracking functions, the DAPs-based one and the usual one, is also implemented. In the system, the "availability" means the value of each aircraft parameter can be obtained continuously, and the "certification" means the obtained value of each aircraft parameter is valid. These two conditions are both important to achieve high accuracy. If the DAPs data has any problem, the IMM filter can provide the output of the system as a usual one in the case of the absence of DAPs. Therefore, for comparing with the usual one, we select a flight with valid DAPs data to evaluate the proposal in the computer simulation.

The paper is structured as follows. In the next section, the mode S and DAPs concepts are described. In Sect. 3, the DAPs-based IMM tracking system is presented. The results of computer simulations and practical experiments are shown and discussed in Sect. 4, and the paper is concluded in Sect. 5.

2 Mode S Enhanced Surveillance

2.1 Mode S Radar

The SSR Mode S includes two elements: an interrogative ground station (GS) and a transponder on board the aircraft. Each GS has its own interrogation code which permits the configuration of the target answers, as they know who is interrogating them. Currently, most Mode S radars are using the multisite surveillance protocol [9]. With this protocol, the GS takes two steps to achieve aircraft surveillance.

First, the GS is searching for and acquiring targets with all-call interrogation. Next, after acquisition, the GS starts tracking targets with aircraft information by using roll-call interrogation. The roll-call interrogation contains the destination address.

The 1090 MHz Mode S extended squitter (1090ES) is also available to periodically broadcast the aircraft information through the Mode S transponder. However, most DAPs cannot be obtained by 1090ES, such as roll angle and track angle rate.

2.2 Downlink Aircraft Parameters

The differences between ELS and EHS fall on uplink and also downlink data transmissions. From a tracking point of view, the most important aspect of ELS is the height measurement improvement (measured in 25 ft steps), which allows better vertical tracking.

The Mode S EHS delivers further information. Thus, an aircraft can send to the ground station several Comm-B Data Selector (BDS) registers with flight information. The decision of what BDSs are sent depends on the local configuration and also on the ATC authority. The ones that seem to be more useful from an air traffic control point of view are shown in Table 1.

Some parameters are for display to the controller, known as controller access parameters (CAPs), and others are for ATM system function enhancement, known as system access parameters (SAPs) [10].

The intention DAPs have an operational utility and are also useful for a collision detection system. For the tracking function, the useful parameters are the vector state DAPs, which tell the ground station how the aircraft is moving at that moment. Even so, not all of them are useful for the tracking function. As our task is to track

Table 1 Downlink aircraft parameters

Register	Parameters
BDS 4.0:	Selected altitude
BDS 5.0:	Roll angle
	Track angle rate (or true airspeed)
	True track angle
	Ground speed
BDS 6.0:	Magnetic heading
	Indicated airspeed or Mach number
	Vertical rate
	True airspeed (if track angle rate is not available)

aircraft in a ground-fixed coordinate system, the most suitable DAPs are the ones referred to in the ground system (BDS 5.0). Four parameters are directly related to the ground-based tracking system: roll angle, track angle rate, true track angle, and ground speed [10, 11]. In this paper, we use the roll angle and track angle rate to detect the turning motion and the ground speed to detect the linear motion with acceleration.

The rest of the parameters referring to the horizontal movement depend on the wind direction and speed, which are useless due to the difficulty of knowing the relationship between them and the ones referred to in the ground system, when they are fused at the ground station.

3 High-Accuracy Tracking System

Air traffic surveillance systems are required to obtain updated and detailed information from aircraft in order to provide safe and efficient air traffic control (ATC) services. In general, civilian aircraft motion can be broadly divided into a uniform motion phase and a maneuver phase. The former refers to the straight and level flight with a constant speed and heading, and the latter refers to acceleration and turning. Especially, the turning motion always arises after takeoff and before landing, and during this period most aircrafts are in the airspace with high density. To satisfy the safety requirements, a good tracking algorithm must provide accurate and reliable estimates of the aircraft state for both modes even in measurement errors and outlier situations.

3.1 Motion Models

The horizontal model and the vertical model are treated separately, due to the fact that the target motions in the horizontal plane and the vertical plane are comparatively independent. In this paper, we only consider the motion models used in the IMM filter in the horizontal plane. The uniform motion can be described by a second-order kinematic (constant velocity) model. The maneuver motion can be described by a third-order kinematic (constant acceleration) model and a coordinated turn model [4, 7]. At each time k, the aircraft's continuous-state dynamics are described by a difference equation defined in the discrete time:

$$x(k) = F_j x(k-1) + G_j w(k-1) \tag{1}$$

where (F_j, G_j) are system matrices with appropriate dimensions, corresponding to each flight mode j, and w is a zero-mean Gaussian white noise used to model accelerations with an appropriate covariance Q_j, which is a design parameter. The three motion models are given as follows.

(1) Constant Velocity Model

The state vector corresponding to the constant velocity model is defined as

$$x = [x\ \dot{x}\ y\ \dot{y}\ ]^T \tag{2}$$

with x and y denoting the orthogonal coordinates of the horizontal plane. And the discrete-time state equation is given by

$$x(k) = \begin{bmatrix} 1 & T & 0 & 0 \\ 0 & 1 & 0 & 0 \\ 0 & 0 & 1 & T \\ 0 & 0 & 0 & 1 \end{bmatrix} x(k-1) + \begin{bmatrix} \frac{T^2}{2} & 0 \\ T & 0 \\ 0 & \frac{T^2}{2} \\ 0 & T \end{bmatrix} w(k-1) \tag{3}$$

where T is the sampling interval.

(2) Constant Acceleration Model

The state vector corresponding to the constant acceleration model is defined as

$$x = [x\ \dot{x}\ \ddot{x}\ y\ \dot{y}\ \ddot{y}\ ]^T \tag{4}$$

The state equation is given by

$$x(k) = \begin{bmatrix} 1 & T & 0 & 0 & \frac{T^2}{2} & 0 \\ 0 & 1 & 0 & 0 & T & 0 \\ 0 & 0 & 1 & T & 0 & \frac{T^2}{2} \\ 0 & 0 & 0 & 1 & 0 & T \\ 0 & 0 & 0 & 0 & 1 & 0 \\ 0 & 0 & 0 & 0 & 0 & 1 \end{bmatrix} x(k-1) + \begin{bmatrix} \frac{T^2}{4} & 0 \\ \frac{T}{2} & 0 \\ 0 & \frac{T^2}{4} \\ 0 & \frac{T}{2} \\ 1 & 0 \\ 0 & 1 \end{bmatrix} w(k-1) \tag{5}$$

(3) Coordinated Turn Model

A coordinated turn is a turn with a constant turn rate (rate of angle change) and a constant speed. Although the actual turning of a civilian aircraft is not exactly coordinated since the ground speed is the airspeed plus the wind speed, the kinematic behavior of the aircraft during the turn is suitably described by the coordinated turn model plus a fairly small noise representing the modeling error. This will be referred to as the nearly coordinated turn model in the sequel. Such a model is necessarily nonlinear if the turn rate is not a known constant. The state vector corresponding to this model is

$$x = [x\ \dot{x}\ y\ \dot{y}\ \omega]^T \tag{6}$$

where ω is the tum rate. The coordinated tum model is then given by

$$x(k) = \begin{bmatrix} 1 & \frac{\sin(\omega(k)T)}{\omega(k)} & 0 & -\frac{1-\cos(\omega(k)T)}{\omega(k)} & 0 \\ 0 & \cos(\omega(k)T) & 0 & -\sin(\omega(k)T) & 0 \\ 0 & \frac{1-\cos(\omega(k)T)}{\omega(k)} & 1 & \frac{\sin(\omega(k)T)}{\omega(k)} & 0 \\ 0 & \sin(\omega(k)T) & 0 & \cos(\omega(k)T) & 0 \\ 0 & 0 & 0 & 0 & 1 \end{bmatrix} x(k-1) + \begin{bmatrix} \frac{T^2}{2} & 0 & 0 \\ T & 0 & 0 \\ 0 & \frac{T^2}{2} & 0 \\ 0 & T & 0 \\ 0 & 0 & T \end{bmatrix} w(k-1) \tag{7}$$

Note that the process noise ν has in general different noise statistics to reflect different modeling errors.

3.2 *IMM Estimator*

One computational cycle of IMM filter consists of four major steps: interaction (mixing), filtering, probability update, and combination [4, 7]. At each time, the initial condition for the filter matched to a certain mode is obtained by mixing the state estimates of all filters at the previous time under the assumption that this particular mode is in effect at the current time. This is followed by a regular filtering (prediction and update) step, performed in parallel for each mode. Then, the mixing and model probabilities are updated by using the likelihood function. Finally, a combination (weighted sum) of the updated state estimates of all filters yields the state estimate. The probability of a mode being in effect plays a key role in the weighting of the mixing and the combination of states and covariances.

(1) Interaction

The motion model j is specified by the state space matrices F_j, G_j, and Q_j. The input, considered as an initial condition, to the filter matched to model j is obtained by mixing the estimates of all filters at the previous time $k-1$ under the assumption that model j is in effect at the present time k. First, the mixing probability $\mu_{i|j}(k-1|k-1)$ is computed as follows:

$$\mu_{i|j}(k-1|k-1) = p_{ij}\mu_i(k-1)/c_j \tag{8}$$

where p_{ij} is the mode transition probability and c_j is a normalizing constant:

$$c_j = \sum_{i=1} p_{ij}\mu_i(k-1) \tag{9}$$

The mode transition probabilities, indicated as assumed to be known, are actually estimator design parameters to be selected in the design process of the algorithm [4].

The mixed initial condition and covariance for the filter matched to model j at time k are

$$\hat{x}_j^o(k-1|k-1) = \sum_{i=1} \hat{x}_i(k-1|k-1)\,\mu_{i|j}(k-1) \tag{10}$$

$$P_j^o(k-1|k-1) = \sum_{i=1} \left[P_i(k-1|k-1) + V_{ij}\right]\mu_{i|j}(k-1) \tag{11}$$

$$V_{ij} = \left[\hat{x}_i(k-1|k-1) - \hat{x}_j^o(k-1|k-1)\right] \times \left[\hat{x}_i(k-1|k-1) - \hat{x}_j^o(k-1|k-1)\right]^T \tag{12}$$

where V_{ij} models increased uncertainty due to disagreement between the model estimates.

(2) Filtering

This step is the same as a single Kalman filtering algorithm. The only difference is that it is to be performed in parallel for each mode. The predicted state estimate for the *jth* filter is

$$\hat{x}_j(k|k-1) = F_j\hat{x}_j^o(k-1|k-1) + G_jw_j(k-1) \tag{13}$$

The predicted covariance associated with the above predicted state estimate is

$$P_j(k|k-1) = F_jP_j^o(k-1|k-1)F_j^T + G_jQ_j(k-1)G_j^T \tag{14}$$

To compute the filter gain, measurement residual (innovation) and residual covariance are defined as

$$\upsilon_j(k) = z(k) - H_j\hat{x}_j(k|k-1) \tag{15}$$

$$S_j(k) = H_jP_j(k|k-1)H_j^T + R_j \tag{16}$$

where $z(k)$ is the measurement at time k, H_j is the measurement matrix of mode j and R_j is the covariance of measurement noise of mode j.

The filter gain is given by

$$K_j(k) = P_j(k|k-1) \times H_j^T S_j(k)^{-1} \tag{17}$$

Then the state estimate and its covariance are written as

$$\hat{x}_j(k|k) = \hat{x}_j(k|k-1) + K_j(k)v_j \tag{18}$$

$$P_j(k|k) = P_j(k|k-1) - K_j(k)S_j(k)K_j(k)^T \tag{19}$$

(3) Probability Update

After each model has been updated in the previous step, new mode probabilities, $\mu_j(k)$, are computed by using the likelihood function $\Lambda_j(k)$, which is defined as

$$\Lambda_j(k) = \frac{1}{\sqrt{|(2\pi)\, S_j(k)|}} \exp\left[-\frac{1}{2} v_j(k)^T S_j(k)^{-1} v_j(k)\right] \tag{20}$$

$$\mu_j(k) = \frac{c_j \Lambda_j(k)}{\sum\limits_{j=1} c_j \Lambda_j(k)} \tag{21}$$

(4) Combination

This step yields the overall state estimate as the probabilistically weighted sum of the updated state estimates of all filters.

$$\hat{x}(k|k) = \sum_{j=1} \hat{x}_j(k|k)\, \mu_j(k) \tag{22}$$

$$P(k|k) = \sum_{j=1} \left\{P_j(k|k) + [\hat{x}_j(k|k) - \hat{x}(k|k)] \times [\hat{x}_j(k|k) - \hat{x}(k|k)]^T\right\} \mu_j(k) \tag{23}$$

3.3 System Architecture

(1) Data Field

Figure 1 represents the DAPs-based tracking system architecture which consists of a Mode S radar with 10 s update rate, a tracker, a DAPs monitor, and a position monitor. The nodes in the system are connected through data field (DF) that serves as

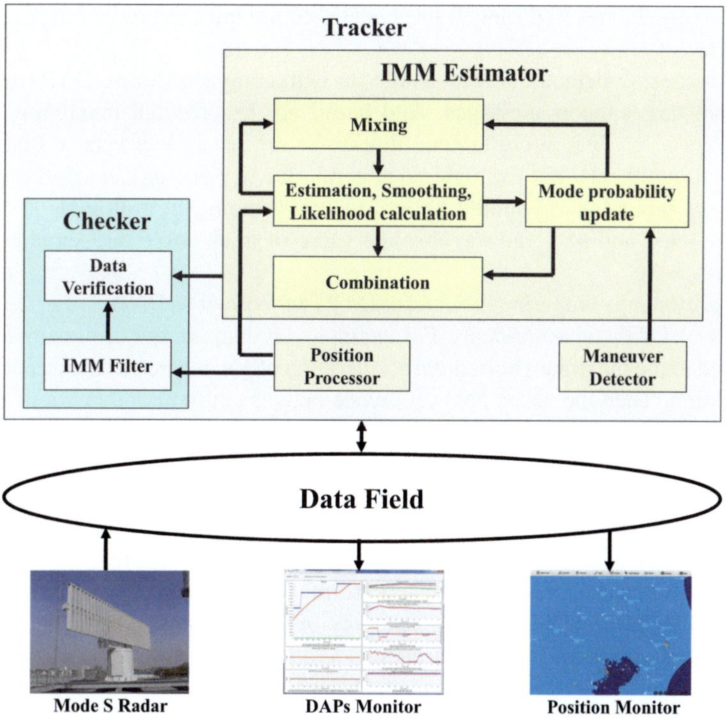

Fig. 1 System architecture

a communication medium of coordination between the nodes. It can be a local area network (LAN) or LANs connected by wide area network (WAN). All necessary data is broadcast into the DF, where the data logically circulates in the DF.

(2) Radar and Monitors

The Mode S radar gets targets' information from its own surveillance area. Then it sends the target report message which contains the position, identification number, time, and DAPs information to the DF.

The DAPs monitor stores the DAPs information of each aircraft and shows the values of parameters of the selected aircraft in the figures. The position monitor displays and updates the measurement, smoothing, and predicted positions of each aircraft.

(3) Tracker

The tracker is composed of a position processor, a maneuver detector, and an IMM filter. The position processor is in charge of storing and updating the position information of aircraft. When receiving the target report message, the processor extracts the aircraft position information and converts it from the radar coordinates (range, azimuth, and flight level) to geodetic coordinates (latitude, longitude, and altitude).

The maneuver detector is responsible for extracting and storing DAPs information from target report messages. And based on the potential maximum energy maneuvers of civilian aircraft, some thresholds are set to detect the validation of DAPs measurements. For example, in general, due to passenger comfort considerations, the ground speed should be less than 600 knots, the roll angle should lie between −45° and 45°, and the absolute value of track angle rate should be less than 8°/s [3, 6].

In addition, by comparing the current DAPs with previous DAPs, it can detect the integrity of DAPs measurements. For example, civilian aircraft are assumed to be subjected to the maximum horizontal acceleration of 1 *g*, where *g* is the gravitational acceleration. Then the acceptable difference between current and previous ground speed should be less than 190 knots in the 10 s radar update period.

Moreover, the consistence of interrelated DAPs is also taken into account. For example, the track angle rate has the same variation trend as the roll angle. The accuracy of the detection of turning motion can be improved, if we consider both roll angle and track angle rate at the same time. Although this is a simple method, it has demonstrated a satisfactory and stable performance during practical experiences.

For assuring the validation and accuracy of DAPs data and outputs of DAPs-based estimator, not only predetection of DAPs measurements is executed, but also the comparison between the outputs of DAPs-based estimator and standard IMM filter is performed. If there is an integrity problem in the DAPs measurements, the default probability update of IMM filter without DAPs data will be applied. And if there is an integrity problem in the outputs of DAPs-based estimator, the results of the standard IMM filter will be selected as the system's output.

According to the above discussion, the following equations are applied to detect the motion model changes of the aircraft based on a set of thresholds.

$$C_a = \begin{cases} 1 & a_1 < \left|V_{gs}(k) - V_{gs}(k-1)\right| < a_2 \\ 0 & \text{otherwise} \end{cases} \tag{24}$$

$$C_t = \begin{cases} 1 & b_1 < |R(k-1) \cdot W(k-1)| < b_2 \\ 0 & \text{otherwise} \end{cases} \tag{25}$$

where $V_{gs}(k)$ is ground speed at time *k*, *R(k-1)* is roll angle at time *k-1*, and *W(k-1)* is track angle rate at time *k-1*. C_a is defined to detect the acceleration, and C_t is defined to detect the coordinated turn.

Based on the analysis of measurement noise and real application of civilian aircraft [11], within 10 s, the acceptable difference between current and previous ground speed should be less than 190 knots, and considering the measurement noise, the thresholds of $a_1=10$ and $a_2=190$ are used to assure validation of acceleration detection. Moreover, due to passenger comfort considerations, the absolute value of roll angle should be less than 45°, and the absolute value of track angle rate should be less than 8°/s. And considering the measurement noise, the thresholds of $b_1=0.02$ and $b_2=360$ are used to guarantee the detection of turning motion.

If the track angle rate is not available, we can calculate the angular velocity which is related to the roll angle and true air speed by the following calculation [11]:

$$W(k-1) = g \cdot \frac{\tan(R(k-1))}{V_{\mathrm{tas}}(k-1)} \tag{26}$$

where g is the gravitational acceleration and $V_{\mathrm{tas}}(k-1)$ is true airspeed at time $k-1$.

In the standard IMM filter [4], the default mode probabilities are used for initialization since it is difficult to predict the motion model of aircraft at the beginning. Therefore, if the default mode probabilities do not match the motion model or the motion model changes during the initial tracking period, the prediction accuracy is degraded. Moreover, during the later tracking period, the mode probabilities become uniform since the difference of residual vectors between modes decreased through the mixing process. As a result, if the motion model changes during the later tracking period, the prediction errors increase because the response of filters is often delayed. To solve this problem, based on the DAPs information, during the initial tracking period, which means a new track is initiated for a new target, the mode probabilities are presented below:

$$\begin{aligned}
&\text{if } (\mathrm{C_t} = 1 \& \mathrm{k} \leq 10)\text{, then}\\
&\quad \mu_1(\mathrm{k}) = P_{ct} P P_{cv}^T\\
&\quad \mu_2(\mathrm{k}) = P_{ct} P P_{ca}^T\\
&\quad \mu_3(\mathrm{k}) = P_{ct} P P_{ct}^T\\
&\text{if } (\mathrm{C_t} = 0 \& \mathrm{C_a} = 1 \& \mathrm{k} \leq 10)\text{, then}\\
&\quad \mu_1(\mathrm{k}) = P_{ca} P P_{cv}^T\\
&\quad \mu_2(\mathrm{k}) = P_{ca} P P_{ca}^T\\
&\quad \mu_3(\mathrm{k}) = P_{ca} P P_{ct}^T\\
&\text{if } (\mathrm{C_t} = 0 \& \mathrm{C_a} = 0 \& \mathrm{k} \leq 10)\text{, then}\\
&\quad \mu_1(\mathrm{k}) = P_{cv} P P_{cv}^T\\
&\quad \mu_2(\mathrm{k}) = P_{ca} P P_{ca}^T\\
&\quad \mu_3(\mathrm{k}) = P_{ca} P P_{ct}^T
\end{aligned}$$

where $\mu_1(k)$, $\mu_2(k)$, and $\mu_3(k)$ are mode probabilities for constant velocity model, constant acceleration model, and coordinated turn model at time k, respectively. P is the mode transition probability matrix, and P_{cv}, P_{ca}, and P_{ct} are the transition matrices for constant velocity model, constant acceleration model, and coordinated turn model, given by

$$P_{cv} = [1 \;\; 0 \;\; 0], P_{ca} = [0 \; 1 \; 0], P_{ct} = [0 \; 0 \; 1]$$

In addition, during later tracking period, which means the difference between mode probabilities is less than a certain threshold, the mode probabilities are revised as follows:

$$
\begin{aligned}
&\text{if } (C_t = 1\&\overline{\mu}(k) < 0.2\&\mu_3\ (k) < 0.5)\text{, then}\\
&\mu_1\ (k) = P_{ct}PP_{cv}^T\\
&\mu_2\ (k) = P_{ct}PP_{ca}^T\\
&\mu_3\ (k) = P_{ct}PP_{ct}^T\\
&\text{if } (C_t = 0\&C_a = 1\&\overline{\mu}(k) < 0.2\&\mu_2\ (k) < 0.5)\text{, then}\\
&\mu_1\ (k) = P_{ca}PP_{cv}^T\\
&\mu_2\ (k) = P_{ca}PP_{ca}^T\\
&\mu_3\ (k) = P_{ca}PP_{ct}^T\\
&\text{if } (C_t = 0\&C_a = 0\&\overline{\mu}(k) < 0.2\&\mu_1\ (k) < 0.5)\text{, then}\\
&\mu_1\ (k) = P_{cv}PP_{cv}^T\\
&\mu_2\ (k) = P_{cv}PP_{ca}^T\\
&\mu_3\ (k) = P_{cv}PP_{ct}^T
\end{aligned}
$$

where $\overline{\mu}(k)$ is the average of differences between mode probabilities

$$\overline{\mu}(k) = (|\mu_1(k) - \mu_2(k)| + |\mu_1(k) - \mu_3(k)| + |\mu_2(k) - \mu_3(k)|)/3 \tag{27}$$

For other conditions, the mode probabilities will be updated by the standard IMM algorithm.

4 Evaluation

4.1 Parameters

The performance of the proposed system is compared with the standard IMM estimator. To obtain the best possible results, the system has to be properly designed to meet the special requirements of the particular sensor. Based on the computer simulation and practical experiments, the design parameters are given in Table 2.

Table 2 Parameters

		Measurement Noise	
Model	Process Noise	Range (m)	Azimuth (deg)
Constant velocity	0.01 g	30	0.06
Constant acceleration	g	30	0.06
Coordinated turn	0.1 g	30	0.06

The model transition probabilities for IMM filter is designed as follows:

$$P = \begin{bmatrix} 0.95 & 0.025 & 0.025 \\ 0.025 & 0.95 & 0.025 \\ 0.025 & 0.025 & 0.95 \end{bmatrix}$$

The two methods are applied to evaluate the performance of the proposed tracking system. One is a real data-based computer simulation. The other is a real-time radar date-based practical experiment. Since the proposed algorithm can use real-time DAPs parameters to accurately detect the current flight motion, it is effective to improve the tracking performance through maneuvers compared with the standard IMM algorithm.

4.2 Results

The first evaluation is a computer simulation. The simulated trajectory is generated based on positions of a real flight trajectory and DAPs information obtained in the practical experience. The smoothing GPS positions are used as the reference (true positions with zero error).

The simulated trajectory starts with a maneuvering segment, followed by the non-maneuvering segment and finally ends with another maneuvering segment, as shown in Fig. 2. The Mode S update rate is 10 s and all DAPs are available.

The performance of the proposed algorithm is compared with the standard IMM estimator through a Monte Carlo simulation. The results shown in Fig. 3 reveal that the proposed system permits to significantly leverage prediction errors through maneuvers in horizontal not only during the initial tracking period (from 10 s

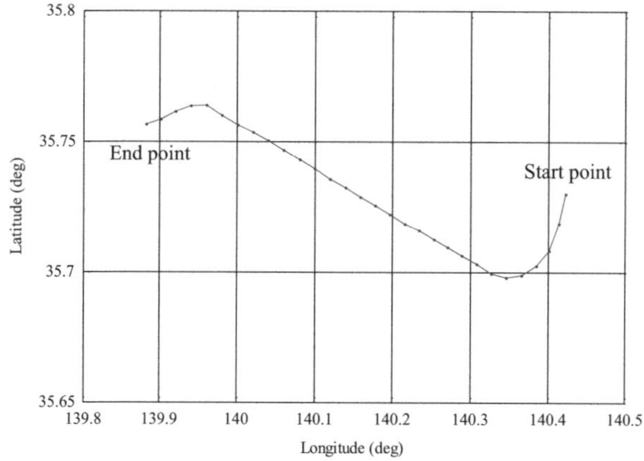

Fig. 2 Trajectory in horizontal

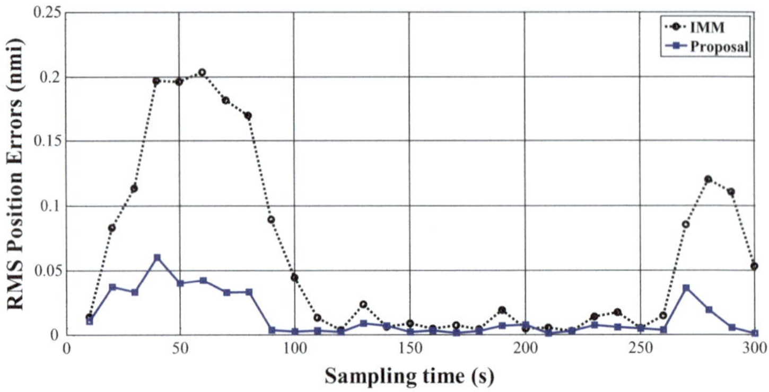

Fig. 3 RMS position errors

Table 3 RMS prediction errors in horizontal

		RMS error (nmi)		
Sampling time (sec)	Measurement points	Proposal	IMM	% Reduction
[0, 100]	62	0.0442	0.1006	56.06
[100, 200]	51	0.0657	0.1022	35.69
[200, 300]	49	0.0433	0.1022	57.63
[300, 400]	37	0.0673	0.1331	49.41
[400, 500]	44	0.0432	0.1111	61.09
[500, 600]	48	0.0427	0.0807	47.16
[600, 700]	64	0.0311	0.05	37.73
[700, 800]	57	0.0317	0.0601	47.33

to 100 s) but also during the later tracking period (from 260 s to 300 s). The average improvement of the prediction error in horizontal is 60 % compared with the standard IMM estimator. During the non-maneuvering periods (from 110 s to 250 s), these two methods have identical performance.

The second evaluation is based on the real-time Mode S radar data that spans a time interval of about 800 s. The aircrafts with DAPs information during the maneuvering periods are selected for comparison. To avoid signal interference, the interrogations are just sent to the selected aircraft to get DAPs information in the practical experiments. Because GPS positions of all flights cannot be obtained, the smoothing measurement positions are used as the reference. In Table 3, the "measurement points" are data points of selected aircraft with DAPs information measured by the Mode S radar and processed by two tracking systems in the sampling time. The actual prediction errors obtained using these two methods are tabulated. These results not only demonstrate the error reduction obtained with the DAPs-based estimator but also indicate the magnitude of the actual errors in a typical ATC scenario. The RMS prediction errors in the horizontal aircraft based on the measurements of maneuvering periods are given in Table 3. The results indicate

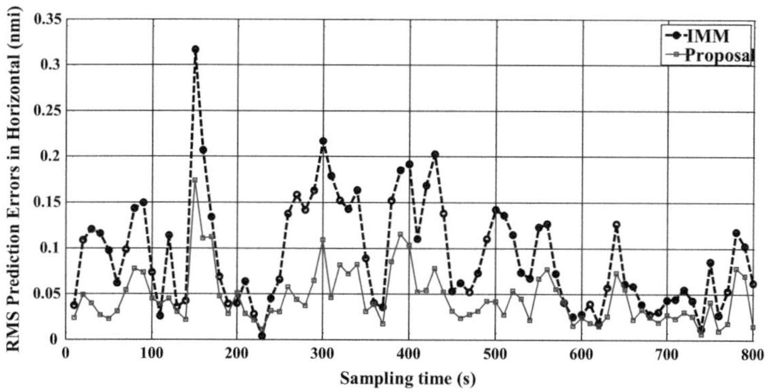

Fig. 4 RMS prediction errors in horizontal

approximately 50 % RMS prediction error reduction in the maneuvering periods for the proposal over the standard IMM estimator. The performance comparison of RMS on horizontal accuracy is shown in Fig. 4. We see that during the maneuvering periods, the proposed system consistently yields lower errors and gets much improvement over the standard IMM estimator.

5 Conclusion

In this paper, the development and evaluation of DAPs-based IMM tracking system for aircraft surveillance are presented. Since the proposed system can use real-time DAPs data to accurately detect the current flight motion and update the mode probabilities, from the results of computer simulation, we can see that it significantly reduces prediction errors through maneuvers not only during the initial tracking period but also during later tracking period compared with the standard IMM-based tracking system without applying DAPs data. The results of practical experiments show a satisfactory and stable performance of the proposed system during tests, and no significant disturbances due to false DAPs measurements and outputs can be observed by the pre- and postvalidation check processes. For satisfying the desired level of accuracy, it is required to further assure the quality of DAPs data and improve the performance during the dynamic changing environment.

References

1. Principles of Mode S Operation and Interrogator Codes, 2nd ed., EUROCONTROL, March 2003.
2. Manual of SSR systems, 2nd ed., ICAO, July 1998.

3. C.C. Lefas, "Using roll-angle measurements to track aircraft maneuvers," IEEE Trans. on Aerospace and Electronic Systems (AES), 6 (1984), pp. 672–681.
4. Y. Bar-Shalom, X.R. Li and T. Kirubarajan, "Estimation with Applications to Tracking and Navigation," Wiley, Hoboken, NJ, 2001.
5. A. Soto, G. de Miguel, J. Besada and J. Garcia, "Robust tracking architecture for mode-s enhanced surveillance," IEEE Proc. of 9th International Conference on Information Fusion, July 2006, pp. 1–8.
6. X.R. Li and Y. Bar-Shalom, "Design of an interacting multiple model algorithm for air traffic control tracking," IEEE Trans. on Control Systems Technology, 3 (1993), pp. 186–194.
7. X.D Lu and T. Koga, "Preliminary study on tracking technologies for hybrid aircraft surveillance," IEICE Technical Report SANE2012-142, January 2013, pp. 55–60.
8. Y. Bar-Shalom and W.D. Blair, "Multitarget - Multisensor Tracking: Applications and Advances," Volume III, Artech House, Norwood, MA, 2000.
9. Aeronautical Telecommunications annex10 vol. IV, 4th ed., ICAO, July 2007.
10. DAP Technical Characteristics for Mode S Enhanced Surveillance, 1st ed., EUROCONTROL, June 1999.
11. R.D. Grappel, G.S. Harris, M.J. Kozar and R.T. Wiken, "Elementary Surveillance (ELS) and Enhanced Surveillance (EHS) Validation via Mode S Secondary Radar Surveillance," Project Report ATC-337, Lincoln Lab., MIT, April 2008.

Experimental Study of Photonic Based Radar for FOD Detection Systems Using 90 GHz-Band

N. Shibagaki

Abstract This paper describes a concept and evaluation results of a Foreign Object of Debris (FOD) detection radar system using Radio on Fiber technologies. Demands for FOD detection system for airports runway is increasing due to explosive growth of air traffic. Since very short wavelength and wide frequency band allocation, 90 GHz-band Frequency-Modulated Continuous-Wave (FMCW) radar with RoF technology is a good candidate for FOD detection system for runway. We have started a R&D project for 90 GHz-band FMCW radar in 2012 and demonstrated in Sendai Airport using the licensed proto-type radar system. Our demonstration system consists of two radar transmitter/receiver unit with rotating antenna and one control unit with signal generation and data analysis capability. These radar unit and control unit are connected with RoF technology. Precision FMCW signal can be shared and transmitted through optical fiber. On Sendai airport runways, we have demonstrated sample FOD (one inch metallic cylinder, etc.) detection from150 m distance range.

Keywords Millimeter-wave • RoF • FOD • Radar • FMCW

N. Shibagaki (✉)
Information & Telecommunication System Company, HITACHI Ltd., Tokyo, Japan
e-mail: nobuhiko.shibagaki.qr@hitachi.com

Electronic Navigation Research Institute, *Air Traffic Management and Systems II*,
Lecture Notes in Electrical Engineering 420, DOI 10.1007/978-4-431-56423-2_13

1 Introduction

The field of microwave photonics [1] has been growing recently and is considered as a disruptive technology for microwave and millimeter-wave applications. In this report, we have chosen 92–100 GHz band for radar frequency. The frequency band is already allocated for Radio Location Service (RLS) by ITU-R. Wide frequency range and small wave length can realize high range resolution radar system. Therefore Foreign Object of Debris (FOD) detection system for runway at the airport is a good candidate for applying the microwave photonics technology with millimeter wave signal. This paper will describe recent studies in Radio-over-Fiber (RoF) based radar system for FOD detection application. [2–6]

Figure 1 illustrates the application for FOD detection system in airports. Optical fiber is used to connect the different radar sub-system along a runway. Radar signal is generated in the centralized control server. The configuration realizes not only high quality signal distribution, but also radar antenna unit simplification and system scalability.

2 90 GHz Band RoF Based RADAR System

Table 1 summarizes system specifications of the RoF based Radar systems. These parameters are achieved using GaAs monolithic integrated circuits, high gain antenna, broadband mixer, multiplier and optical components.

As shown in Table 1, the radar system consists of signal generation unit and the radar unit. Signal generation unit generates Frequency-Modulated Continuous-Wave (FMCW) signal of 10 GHz-band and convert it to optical signal using optical modulator. The optical FMCW signal is transmitted to Radar unit via optical fiber. Figure 2 shows simplified block diagram of a RoF based Radar unit. The optical FMCW signal is converted to RF signal with Photo detector. The 10 GHz-band RF signal can be multiplied to 90 GHz-band RF signal. GaAs monolithic integrated high power amplifier amplifies the multiplied signal up to 100 mW power range to realize 500 m detection range.

The optical FMCW signal is also multiplied and then supplied to a receiver mixer to generate a beat signal with the reflection signal from FOD target. Figure 3 shows an external view of the Radar unit. Tx module (multiplier, BPF and HPA) and Rx module (multiplier, Mix. and LNA) are hermetically packaged and fixtured at back side of IF and peripheral circuit board to realize ideal configuration with wave guide antenna port.

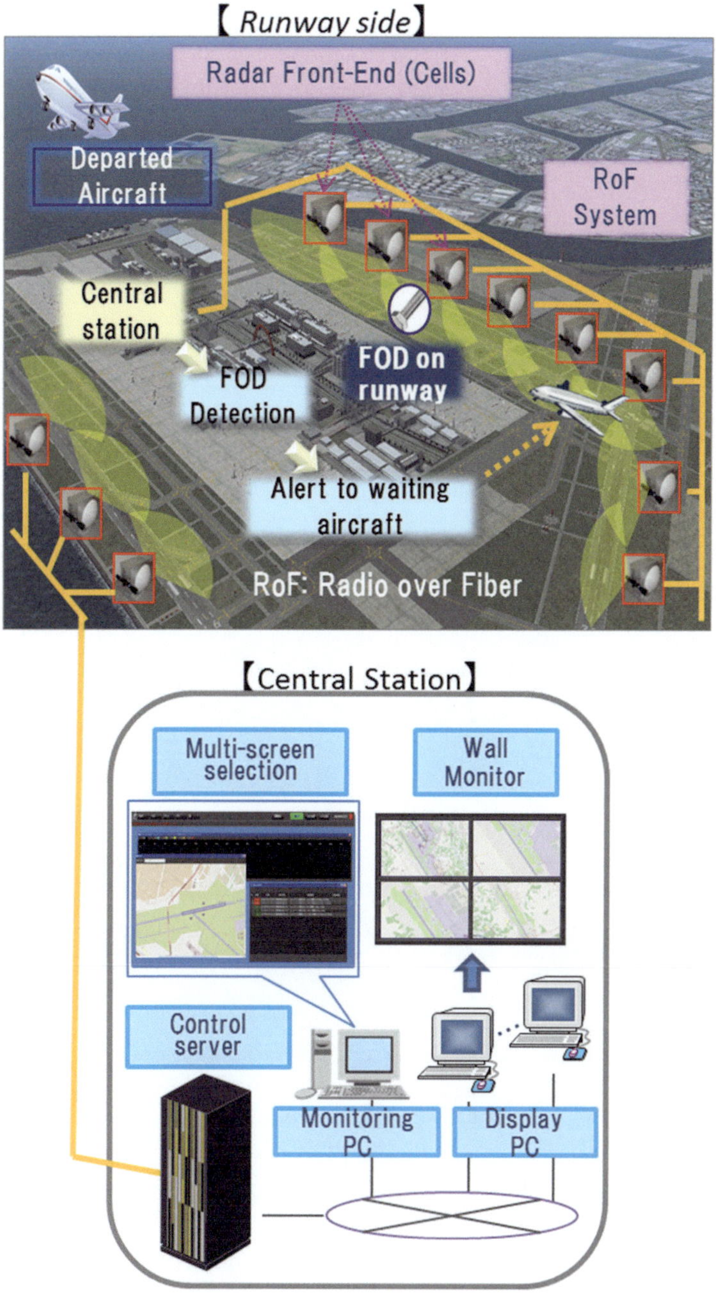

Fig. 1 Schematic view of a RoF based radar system for FOD detection system in runway at airports

Table 1 System specification

Spec./Block	RADAR unit(RF)	Signal generation unit(IF)
Output frequency range(GHz)	90 GHz-band	10 GHz-band
Output power	100 mW	10 mW
Sweep range	92.05∼99.95 GHz	10.227∼11.105 GHz
Sweep frequency	∼1250 Hz	
I/F(Antenna)	WR-10	N/A
Antenna type & FWHM	Offset Parabola	N/A
	Elevation:1.0°	
	Azimuth:1.0°	
Antenna rotation speed	15 rpm	N/A
Detection range	500 m	
Range resolution	5 cm	

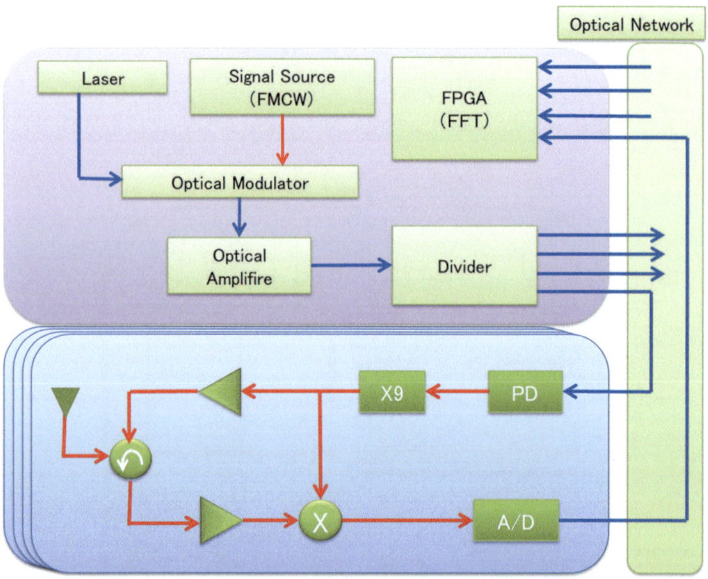

Fig. 2 Block diagram of a RoF based radar system

Fig. 3 External view of a RoF based radar unit

3 Characteristics of 90 GHz-Band RF Module

3.1 Characteristics of Tx Modules for High Power Transmission

The extremely wide frequency range can realize good range resolution. For this purpose both Tx and Rx module should have wide band and flat frequency response.

Figure 4 shows the frequency responses of the Tx module. More than 20 dB conversion gain and saturation power of 20 dBm is observed at 92–100 GHz frequency range. Input frequency range is 10 GHz-band for this measurement. These measurement results also indicate that conversion gain variation within the desired frequency range is less than 2 dB.

3.2 Characteristics of Rx Modules for Low Noise Reception

Figures 5 and 6 show frequency response and IF output signal spectrum of the Rx module respectively. Assume a 500 m distance between radar and target, time difference between transmitted signal and reflected signal T=3.34 us (=2×500 m/c, where c is the velocity of light). Since our radar system employs 8 GHz modulation bandwidth and 800 us period, we can get "beat signal" of roughly 70 MHz (=8 GHz/800us/2×T) using single ended mixer. The mixer has −10 dB gain and flat frequency characteristic in the operating frequency range. The power level of

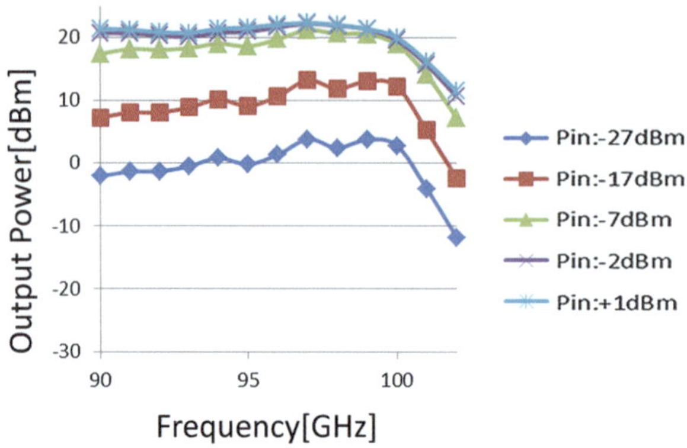

Fig. 4 Frequency response of Tx module

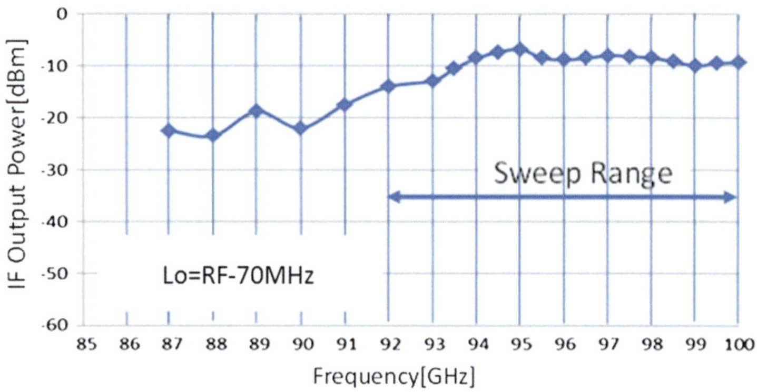

Fig. 5 Frequency response of Rx module

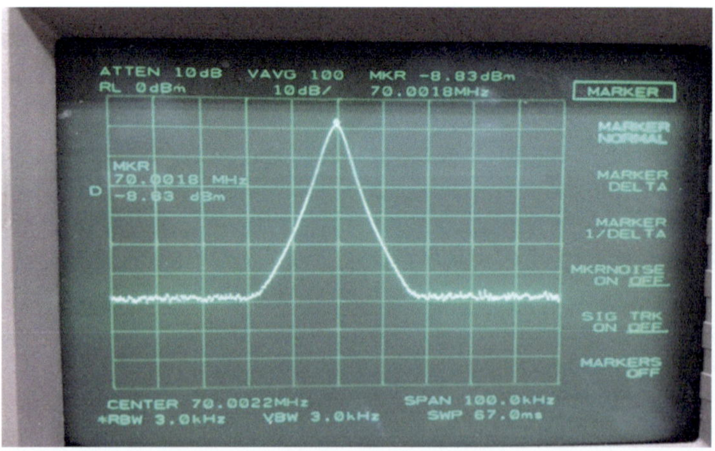

Fig. 6 Frequency spectrum of IF output signal

RF input for the measurement is −50 dBm so that the conversion gain of the Rx module is 40 dB in the frequency sweep range. Since the S/N ratio of the IF output signal is more than 60 dB for −50 dBm input RF signal, we can tentatively estimate minimum receiving sensitivity of the Rx module is −110 dBm. Receiver conversion gain and system noise level are vitally important and determine the maximum detection range. According to the FAA Advisory Circular for Airport Foreign of Debris Detection system [3], Airport Foreign of Debris Detection systems have to detect metallic cylindrical target with 3 cm height and 3.8 cm diameter. Considering the Advisory Circular, we have set our system goal is to detect objects of −20 dBsm Radar Cross Section (RCS) at a 500 m distance. The −20 dBsm of RCS is equivalent to RCS of metallic cylindrical target with 1 inch height and 1 inch diameter. Using radar equation with Tx/Rx antenna gain of 42 dBi, and emission power of 100 mW, we can estimate that receiving signal level from −20 dBsm RCS object at a 500 m distance is −107 dBm. Although the tentatively estimated minimum receiving sensitivity of −110 dBm is approaching the target, we need more noise performance improvement and systematic evaluation for realizing good system margin.

4 System Evaluation

4.1 Detection Range

We have carried out a system evaluation of the RoF based Radar system with our proto-type Radar system. The system noise level of the proto-type radar system is 20–30 dB higher than we describe previous section. Due to this reason, we use +5 dBsm RCS reference target instead of our final target (−20 dBsm RCS) for maximum detection range evaluation. For this evaluation, we employ "Omni-directional dielectric lens and wide angle lens reflector [4]" as reference target. As shown small picture in Fig. 7, we have set-up reference target at a 500 m distance position from the Radar. In the Plan Position Indicator (PPI) image, we can observe not only the reference target, but also utility pole and curb at road-side etc.

4.2 Range Resolution

Theoretical range resolution ΔR of FMCW Radar is simply estimated by equation $\Delta R = c/(2 \cdot f_{BW})$, where c is the velocity of light and f_{BW} is the modulation bandwidth of the Radar system. Since we employ 8 GHz modulation bandwidth, ΔR is theoretically less than 2 cm. However, it is generally difficult to realize the theoretical range resolution with actual hardware/software implementation due to the effect of the spectrum fluctuation and the FFT window.

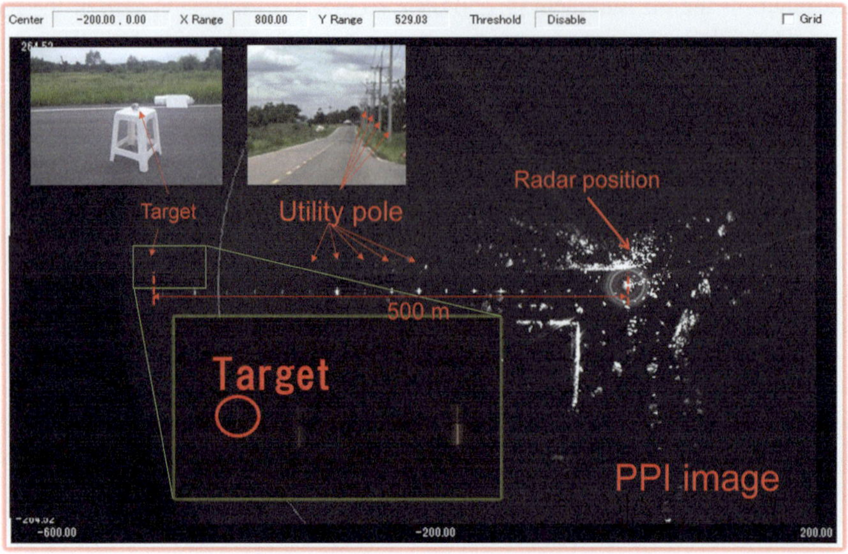

Fig. 7 PPI image of 90 GHz-band RoF based radar

Fig. 8 Photograph of sample FOD set-up

Figure 8 shows two −20 dBsm targets of the range resolution measurement. These targets are located at the position at exactly 60 m away from the Radar. One target is fixed at 60 m and another is located at 60 m+5 cm or 60 m-5 cm.

The measured Radar range profile is shown in Fig. 9.

Red line in Fig. 9 indicates the target position of 59.95 m and 60 m, blue line indicates the target position of 60 m and 60.05 m. The 60 m peak is observed both case and 59.95 m peak is observed in red line, 60.05 m peak is observed blue line.

The measured 5 cm range resolution at 60 m distance range is a reasonable value compared with the theoretical value.

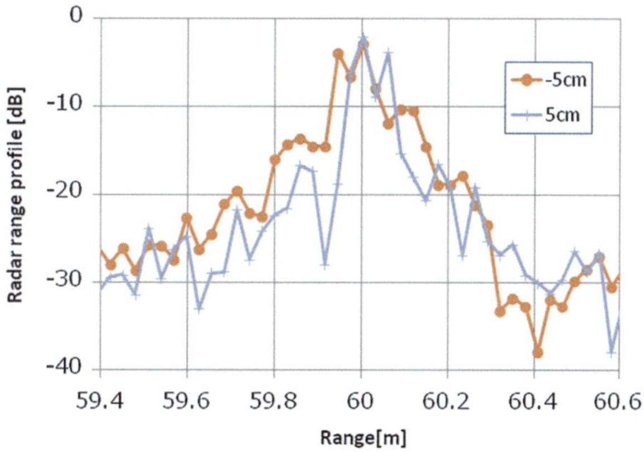

Fig. 9 Radar range profile at a 60 m distance

5 Conclusion

We have demonstrated RoF based 90 GHz Radar system for FOD detection system for airport runway. Using integrated monolithic GaAs module, we confirmed the feasibility of reasonable detection range (500 m for +5 dBsm RCS) and range resolution (5 cm separation at 60 m). We are planning a demonstration at Narita Airport runway with four Radar units.

Acknowledgments This study was conducted as a part of a research project entitled “R&D of high-precision imaging technology using 90 GHz-band linear cells”, supported by a Japanese Government funding, “R&D to Expand Radio Frequency resources” of Ministry of Internal Affairs and Communications.

References

1. A.J. Seeds, “Microwave photonics,” IEEE Transaction on Microwave Theory and Techniques, vol. 50, pp. 877–887, 2002
2. S. Futasumori, A. Kohmura, and N. Yonemoto, “Performance measurement of compact and high performance 76 GHz millimeter-wave system for autonomous unmanned helicopters”, IEICE Technical Report., SANE 2011-53, July 2011.
3. N. Yonemoto, A. Kohmura, S. Futatsumori, T. Ueno, and A. Saillard, “Broad band RF module of millimeter wave radar network for airport FOD detection system,” in Proc. Int. Radar Conf.-Surveillance Safer World, Oct. 2009.
4. A. Kohmura, S. Futatsumori, N. Yonemoto, and K. Okada, “Optical fiber connected millimeter-wave radar for FOD detection on runway,” in Proc. 10th Eur. Radar Conf., Oct. 2013, pp. 41–44(2013).
5. A. Kanno and T. Kawanishi, “Broadband Frequency-Modulated Continuous-Wave Signal Generation by Optical Modulation Technique,” IEEE/OSA J. Lightw. Technol., vol. 32, no. 20, pp. 3566–3572(2014)
6. S. Futasumori, K. Morioka, A. Kohmura, K. Okada and N. Yonemoto, “Evaluation of High-speed FMCW Signal Generation and Processing for Optically-Connected Distributed-Type

Millimeter-Wave Radar", Proceedings of the 2015 International Symposium on Antenna and Propagation (ASAP2015), pp. 903–904, Nov. 2015.
7. FAA Advisory Circular for Airport Foreign of Debris Detection system #FAA AC150/5220-24. http://www.faa.gov/documentLibrary/media/Advisory_Circular/150_5220_24.pdf
8. N. Yonemoto and M. Matsuzaki, "An Omni-directional Lens Reflector for Millimeter Wave and Light", Proceedings of International Radar Symposium 2007(IRS 2007), pp. 575–578, Cologne, Germany, September, 2007

Printed by Libri Plureos GmbH
in Hamburg, Germany